D0577516

Inside Animal Hearts and Minds

Bears That Count, Goats That Surf, and Other True Stories of Animal Intelligence and Emotion

Belinda Recio
Foreword by Jonathan Balcombe

Skyhorse Publishing

Skyhorse Publishing books may be purchased in bulk at special discounts for sales promotion, corporate gifts, fund-raising, or educational purposes. Special editions can also be created to specifications. For details, contact the Special Sales Department, Skyhorse Publishing, 307 West 36th Street, 11th Floor, New York, NY 10018 or info@skyhorsepublishing.com.

Visit our website at www.skyhorsepublishing.com.

10 9 8 7 6 5

Library of Congress Cataloging-in-Publication Data is available on file.

Cover design by Jane Sheppard
Cover photo credit: iStockphoto

Interior book design by Belinda Recio
Interior photo credit: iStockphoto, except where specified

Print ISBN: 978–1–5107–1894–4
Ebook ISBN: 978–1–5107–1895–1

Printed in the United States of America

CONTENTS

PART ONE: HEART

PART TWO: MIND

The spirit wanders, comes now here, now there, and occupies whatever frame it pleases. From beasts it passes into human bodies, and from our bodies into beasts, but never perishes.

—Ovid, *Metamorphoses*

Foreword

I often tell audiences that these are exciting times to be an ethologist (a specialist in animal behavior), for we are living in an unprecedented era of discovery of the inner lives of animals. I sometimes add, with some pride, that I may be the first person with the phrase "Animal Sentience" in their job title. (For the uninitiated, "sentience" is the capacity to feel.) Since 2014, I have been Director of Animal Sentience with the Washington, DC-based Humane Society Institute for Science and Policy.

It was in that role, in March 2014, that my colleagues and I staged a two-day conference in DC titled "The Science of Animal Thinking and Emotion." The meeting brought together luminaries in the fields of ethology and sentience from North America and Europe. We learned about such things as tool use and cooperation by crocodiles, the flexible language of prairie dogs, and the way a dog's brain lights up when greeting her beloved guardian. That last example points to two other aspects of progress in our understanding and appreciation of animals' experiences: 1) the advancement of technologies that allow us to explore previously inaccessible facets of animals' inner worlds, and 2) approaches that treat animals as autonomous beings with preferences and moods, rather than as dumb things to be bent to the human will. The dogs used in Greg Berns's innovative research at Emory University are people's companion animals, trained through positive reinforcement (read: food treats) to remain still in an fMRI (functional magnetic resonance imaging) machine. They participate voluntarily and they go home after a two-hour session in the lab.

Dogs happen to be the new darlings of animal cognition research, and one has to wonder why it has taken so long for scientists to focus on an animal with which we have such a strong affinity borne of thousands of years of cohabitation and communication. Within months of our conference, I had the opportunity to visit the Clever Dog Lab in Vienna, Austria, where I saw, among other methods, dogs learning to rest their chins on a padded platform so that their eye movements and tail positions could be monitored with highly sensitive cameras while they looked at various projected images and videos. Methods like this reveal that dogs glance first to the left when they look at a human face. Why do that, you may be wondering? Apparently, to get a useful quick read of our emotional state, which is better revealed on that side of our faces by our bilateral brains. We do the same thing when we glance at human faces, even though we, and most likely they, are not aware we are doing it. The chin-rest apparatus has also fostered the discovery that dogs wag their tails more to the left when they are feeling anxious, as when confronted with a photograph of an unfriendly looking, unfamiliar dog. That might be a useful bit of information for a dog who is about to move in for that sought-after sniff just below the tail.

Scientific fascination with animal thinking and feeling seems natural enough, but it represents a sea change, given that for much of the twentieth century the idea of animals having conscious minds was roundly rejected and deemed unsuitable for study.

It's one thing that scientists are getting on board with perceiving animals as they are—thinking, feeling beings. But what about society? One needs look no further than the bookshelf. Since Jeffrey Moussaieff Masson and Susan McCarthy published the fabulously popular *When Elephants Weep: The Emotional Lives of Animals*, in 1995, the field has been expanding fast. In just the last two years, *Beyond Words* by Carl Safina, *Soul of an Octopus* by Sy Montgomery, *The Genius of Birds* by Jennifer Ackerman, *Being a Dog* by Alexandra Horowitz, *Are We Smart Enough to Know How Smart Animals Are?* by Frans de Waal, *Voices of the Ocean* (dolphins) by Susan Casey, *The Hidden Lives of Owls* by Leigh Calvez, and my own *What a Fish Knows* have all made the *New York Times* best-seller list.

It is my hope that the book you are now holding will also ignite the public imagination. The pace of new discoveries warrants new books to bring it to a hungry public that generally avoids the rarefied, jargon-steeped journals where this stuff first, and often only, gets published. Perhaps some animals don't need a public relations plug, but who wouldn't be surprised to learn that dolphins play with breaching humpback whales by riding up on the leviathans' backs then sliding down or that young chimpanzees play games of make-believe? Other species badly need some good press, and so I was delighted to learn that rattlesnakes are more sociable than we thought and that they babysit the young of other rattlesnake moms. And that their cousins, the crocodilians, have recently been found to engage in joyous behaviors unbefitting their stereotype, including sliding down slopes, surfing on waves, and playing games of chase.

It is tempting to share more of these gems here, but better to let you uncover them yourself. Belinda Recio relates it all in crisp, informative vignettes accompanied with photographs, and an occasional touch of humor.

I leave you with one caution as you read this book, and others that celebrate the mental faculties of our animal brethren: we should be mindful not to attach too much importance to animal intelligence. What matters most is that animals feel, and that they experience their lives no less intensely than we do ours. Nor should we think of humans as the pinnacle against which all others are measured. Belinda Recio clearly knows this, and many of her examples reminded me of the folly of this thinking. As Henry Beston so beautifully expressed in his immortal characterization of animals as "other nations," it is not just when we recognize how they may resemble us, but in how they differ from us in their approaches to life that our own hearts and minds may be most uplifted.

—Jonathan Balcombe

acknowledgments

I am profoundly indebted to all the scientists, animal caregivers, and animal advocacy organizations whose research, observations, and humane education programs inspired this book and are changing how we understand and treat animals.

Special thanks to those scientists and researchers who reviewed my drafts and helped me to better present their research, including Irene Pepperberg, Con Slobodchikoff, Jennifer Vonk, and James Wood.

Much gratitude and appreciation to Jonathan Balcombe for writing the foreword.

I am very grateful to *Organic Spa Magazine* for its support of my column, "State of the Ark," particularly to my previous editor Mary Bemis, who invited me to launch the column in 2010, my current editor Rona Berg, and publisher Beverly Maloney Fischback.

Thank you to my editor, Kim Lim, and Skyhorse Publishing for believing in this book.

A warm thank-you to my family and friends who listened to my animal stories, shared in the enchantment of the research, and supported me in various ways.

Special thanks to Eileen London, for her review of the manuscript; and to Joan Parisi Wilcox, for her editorial expertise, sage advice, and steadfast support throughout this project.

My deepest gratitude and immeasurable appreciation to my husband, Ed Blomquist, for our countless conversations about this book, his manuscript review, and all the ways he supports my work.

author's notes

Every effort has been made to credit the scientists, researchers, and writers whose work I present herein. I regret any omission and pledge to correct errors called to my attention in subsequent editions.

Regarding terminology: although humans are animals, I use the term *animal* to refer to nonhuman animals. In order to avoid objectifying animals, instead of *it* I use the pronouns *him* and *her*.

For my mother Maria, who taught me to see the beauty and soul in animals; for my husband Ed, for understanding why I had to write this book; for my dog Spooner, who helps me to remember what's important in life; and for all the other animal hearts and minds that I have had the immense pleasure of knowing and loving.

INTRODUCTION

One day we humans may meet an intelligent being from another world. Hollywood tells us this stranger will come flying down in a spaceship, and will look a bit like us. But maybe it won't be like that. Maybe it will be like this.

—Ryan Reynolds, in *The Whale*

Shooter, the resident elk at the Pocatello Zoo in Idaho, is huge—standing six-feet tall without taking into account his massive antlers, and ten-feet tall if you do. As you can imagine, he attracts a lot of attention, and one day a few zoo staffers watched as he acted strangely. He repeatedly dipped his entire head into his water trough, a decidedly uncharacteristic elk behavior. He certainly wasn't just taking a drink. Then, things got even more curious. Pulling his head from the water, Shooter now dipped his front hooves into the trough and seemed to be rooting around. After a minute or so, he withdrew his hooves and dunked his head again. When his head emerged this time, he had a tiny dripping marmot in his mouth. He carefully released the marmot on the ground and nudged him with one hoof until—recovered from his near drowning—he scampered away.

What was going on? Did Shooter rescue a drowning marmot? That's what one of the zookeepers, Dr. Joy Fox, thinks. She believes Shooter sensed that the little rodent was in distress and decided to help. But on his first try, his antlers prevented him from reaching the marmot. So, Dr. Fox suggests, Shooter used his hooves to nudge the marmot away from the edge of the trough, then dunked his head again and rescued the marmot.

Such behavior suggests that Shooter may have displayed two capacities not usually attributed to animals: the ability to problem solve and empathy. It's clear that Shooter figured out how to get the marmot out of the trough, though no one knows why. Perhaps he just didn't like rodents in his drinking water or perhaps he was trying to help the marmot. Throughout history, most scientists and philosophers would have denied the possibility of altruistic behavior in an elk, especially toward another species. But today, scientists are not only willing to concede that animals have intelligence and emotions, they are also going out and finding the evidence to prove it.

Scientists have shown us that many animals have rich emotional lives: rats laugh when tickled, magpies appear to mourn as they cover their fallen friends with greenery, female humpback whales travel thousands of miles for annual reunions with their gal pals, young chimps play make-believe and pretend that sticks are babies, capuchin monkeys express indignation when treated unfairly, dogs comfort other species in distress, and elk may be altruistic.

Researchers are also amassing astonishing evidence for animal cognition. In fact, many scientists no longer question *whether* animals think; now they are asking *how* animals think. Just as importantly, they are increasingly recognizing that human bias can get in the way of how science evaluates animals' cognitive abilities. In his book *Are We Smart Enough to Know How Smart Animals Are?*, primatologist Frans de Waal suggests: "Instead of making humanity the measure of all things, we need to evaluate other species by what they are. In doing so, I am sure we will discover many magic wells, including some as yet beyond our imagination."

Friendship is good for the heart—and possibly for a species' survival, too. In one group of female humpback whales, those with the most friends gave birth to the most calves.

Discoveries about animal intelligence are revealing nothing short of de Waal's "magic wells." Dung beetles take mental snapshots of the Milky Way and use these celestial maps to navigate. Bears and chickens can count, whales compose songs that rhyme, and ants appear to recognize themselves in mirrors. When apes who can "talk" using signs or lexigrams don't know the word for something, they occasionally invent their own terms, some of which are startlingly creative metaphors, such as calling ketchup "tomato toothpaste." Prairie dogs have a complex vocabulary, with calls that function as parts of speech. Using specific yips and barks, they can communicate information as detailed as, "Watch out for the small human in the blue shirt!"

The examples of animal hearts and minds you will find in this book might sound like the stuff of storybooks, not science. But it *is* science. And it is proving what Charles Darwin suggested more than a century ago: the difference between humans and other animals is "one of degree and not of kind." As I selected examples of animal emotion and cognition for this book, my biggest challenge was not in finding reliable scientific evidence; it was in choosing from so many amazing stories. There is a continual outpouring of astonishing discoveries about animals, almost on a weekly basis. The evidence for their capacity to feel a wide range of emotions and act with insight and intelligence is paradigm-shifting—radically changing the way we view animals, the world, and our place in it.

The great American naturalist Henry Beston described animals as "other nations" that are "living by voices we shall never hear." Beston probably never imagined that scientists would one day give voice to these other nations, but that is exactly what they have begun to do. My mission in writing this book is to help those voices be heard, with the hope that as we learn more about the hearts and minds of animals, we will more readily recognize them as kindred spirits and treat them—and the environments that sustain them—more compassionately.

"They are not brethren, they are not underlings: they are other nations, caught with ourselves in the net of life and time, fellow prisoners of the splendor and travail of the earth."

—Henry Beston

Part One
Heart

The universe is full of magical things, patiently waiting for our wits to grow sharper.

—Eden Phillpotts, *A Shadow Passes*

1. THE JOKE'S ON US

LAUGHTER, HUMOR, AND MISCHIEF

WHISKERED MIRTH

If you want to know the best place to tickle a rat, ask Jaak Panksepp. Although he has been tickling rats for more than twenty years, it didn't take him long to figure out that rats have an especially ticklish spot at the napes of their necks. Panksepp, a neuroscientist who studies the origins of emotions, started tickling rats after he noticed that they were making a high-pitched chirping sound just before they played. It seemed as if they were chirping in cheerful anticipation of the rough and tumble fun to come. Panksepp wondered if this chirping was the rat equivalent of laughing. So, he and his research team started tickling rats.

> Rats will run mazes and press levers—activities most often rewarded with food—just to get a good tickle.

First, they established a strict tickling protocol: using only your right hand, briskly tickle the rats' backs and necks, then gently flip them, tickle their tummies, and release them. During the tickling, Panksepp used specialized audio equipment to enable the research team to hear the rats' chirps, which are high-pitched (fifty kilohertz) and therefore out of the human hearing range.

The rats not only chirped throughout the tickling, but chirped twice as much when tickled as when playing on their own. And they chirped especially intensely when tickled at the napes of their necks, which is how Panksepp discovered the rats' special tickle spot. It turns out that the nape area is where rats tend to play-bite when they wrestle and chase one another, so it may have evolved into an especially ticklish spot.

The rats seemed to love being tickled. In fact, when the researchers stopped tickling them, the rats would chirp and play-bite the tickler's hands, which is how rats engage one another when they want to keep playing. The rats would even run mazes and press levers—activities most often rewarded with food—just to get a good tickle. As the tickling experiment continued, Panksepp discovered that those rats who chirped the most played the most; and chirping rats preferred the company of other rats who chirped. Further, when it came to romance, female rats preferred cheerful, chirping mates to more serious and sober suitors.

When it comes to rats, it isn't misery that loves company, but cheerfulness.

After years of research, Panksepp became increasingly convinced that when rats chirp, they are, in fact, laughing. The rats chirped when they were playing together, when exhibiting bonding behavior with the tickler's hands, and when other rats chirped. These are conditions similar to those that invoke laughter in humans: we laugh when we are having fun, when we feel bonded, and when others laugh. Even a recording of rats chirping made rats chirp, which can't help but call to mind the reason sitcom laugh tracks were invented.

Panksepp and other scientists believe that laughter probably predates language and evolved through play as a way to communicate the emotional state of the participants. Laughter can signal that "things are good between us—let's keep at it," which can increase the duration of socializing. Between rats, laughter—or chirping—makes play last longer. (Between people, laughter doesn't only lengthen the time spent playing together, it also makes conversations last longer.) Panksepp believes that the "neural circuits for laughter exist in very ancient regions of the brain, and ancestral forms of play and laughter existed in other animals eons before we humans came along with our hahahas and verbal repartee."

In other words, if rats have the neural circuits for "laughter," then it is reasonable to assume that they actually are laughing. As the saying goes, if it looks like a duck, swims like a duck, and quacks like a duck, then it probably is a duck—or, in this case, a laughing rat.

However, Panksepp's colleagues were not so sure. Until recently, most scientists agreed with Aristotle, who asserted, "Only the human animal laughs." Along with other emotional capacities, laughter was considered an exclusively human trait, so Panksepp's hypothesis was initially greeted with skepticism. But after many years of rigorous data collection and analysis, his rat laughter research is not only gaining acceptance, it's also inspiring other scientists to pursue similar areas of study. Panksepp gets the last laugh (or chirp) after all.

Slapstick and Mischief

Like rats who chirp in response to pleasurable physical contact—such as tickling, wrestling, chasing, and other forms of play—chimpanzees, bonobos, gorillas, and orangutans also produce laughter-like vocalizations. Similar to humans, chimps open their mouths, expose their teeth, and make a noise. Instead of "hahaha," however, their "laughter" sounds breathy, almost like they are panting. Chimps, like rats and humans, find laughter contagious and often laugh simply in response to other chimps laughing.

Chimps love a good practical joke as much as people do.

Captive chimps are known for laughing at slapstick humor. They seem to find the clumsiness and gullibility of others as funny as people do. At zoos and research facilities, chimps often amuse themselves at the expense of their caregivers or visitors. For example, Georgia, a captive chimp at the Yerkes Primate Research Center, has a favorite routine in which she spits water at unsuspecting people. After getting a drink, she keeps some of the water in her mouth, sometimes for a surprisingly long time. She walks around, acting nonchalant, as if is she simply going about her business. But as soon as a target is in range, Georgia blasts the water at the unwitting butt of her joke. She and the other chimps burst into shrieks of laughter, often laughing so hard they fall over and roll around on the ground.

Humor is often defined as seeing improbable connections or connecting illogical things. This is what Koko, the famous "talking gorilla," does when

Captive gorillas appear to enjoy a good prank, too. One gorilla at the ZSL London Zoo would poke at strangers with a stick and then look the other way, acting just as a child might after pulling the same stunt.

she wants to crack a joke. Koko knows more than one thousand words in American Sign Language and understands roughly two thousand words of spoken English. Once, to tease one of her teachers, Koko signed that she would like to drink through her ear. Koko knows that you can only drink through your mouth, and she knows and regularly uses the correct signs for mouth, ears, and nose, so she wasn't "misspeaking." Instead, her trainer, Penny Patterson, believes Koko was intentionally joking around—playing with the humorous impact of incongruity. At other times, Koko's humor was definitely not so subtle. Once, she tied Patterson's shoelaces together and signed, "Chase!"

> Dolphins and parrots are also notorious pranksters, often teasing or making jokes at another's expense just for fun.

Dolphins may not share Koko's sign language skills, but when it comes to mischievous humor, these aquatic comics can be as naughty as gorillas or chimps. It is common knowledge among animal behaviorists and people who live near dolphins that they will occasionally "prank" pelicans, seagulls, and cormorants by swimming silently behind the unsuspecting birds, then plunging forward and pulling their tail feathers. Other times, they swim underneath and rise up to toss the birds into the air. Dolphins tease fish by offering them bait and snatching it away at the last minute. They even play catch by tossing turtles, octopuses, and fish (the ones they don't eat) back and forth to one another. As far as scientists can tell, dolphins do these things seemingly for no purpose other than their own amusement.

Parrots are notorious pranksters, too, known for teasing dogs by imitating human voice commands, dropping objects on people and pets, and engaging

Like humans, dolphins are socially complex animals. On one hand, they are gregarious, affectionate, and altruistic; on the other, they are scheming, mischievous, and sometimes even naughty, as when they amuse themselves at another's expense.

in other antics. Sometimes they even demonstrate an edgy and dark sense of humor. In his book *The Parrot's Lament and Other True Tales of Animal Intrigue, Intelligence, and Ingenuity*, Eugene Linden recounts a few stories about animal psychologist Sally Blanchard and her two female parrots, Bongo Marie and Paco. Unfortunately, Bongo Marie did not like Paco, and she found humorous ways to show that dislike. For example, Bongo Marie would taunt Paco by repeatedly calling out her name. When Paco eventually responded, Bongo Marie would scold her with a stern, "Be quiet!"

Once, when Blanchard was cooking a Cornish game hen and pulled it out of the oven, Bongo Marie cried out an emotional, "Oh, no! Paco!" Blanchard reassured Bongo Marie that the roasted hen was not Paco: "See? Paco is right over there." Bongo Marie's response was a disappointed, "Oh, no," after which she laughed maniacally. Because the display of humor was so sophisticated, skeptics believed that Blanchard embellished this story, but people who have parrots and who regularly witness their level of intelligence and awareness have less trouble believing it.

People who live with parrots tell all kinds of stories about the ways these brainy birds trick and tease their family members.

Grin and Bear It

Often referred to as a bear whisperer, Else Poulsen was a biologist who devoted her life to the care and study of bears. During her years spent in zoos and wildlife sanctuaries, Poulsen observed what she described as a "bear sense of humor." When people joke—or even just anticipate a joke—they smile. So do bears. In fact, they smile pretty much the same way that humans do, turning the sides of their mouths upwards. Sometimes, those bear smiles give way to laughter. According to Poulsen, bears laugh "like gorillas, with their mouth wide open, while silently moving or bobbing their head about, resembling one of Jim Henson's Muppets."

During her time as a zookeeper at the Calgary Zoo, Poulsen worked with a pair of grizzly bears, a female named Khutzy, who was raised in captivity, and a male named Skoki, who had been relocated from a national park. Skoki was a "problem bear" in the park, known for stealing food from tourists. When Skoki decided to walk through the front door of a local bakery to see what they had to offer, park officials decided it was time for Skoki to find a new home.

Despite their different backgrounds, the two grizzlies became fast friends and played together like cubs. According to Poulsen, their play sessions involved lots of smiling and open-mouth laughing. One day she observed what appeared to be a kind of private joke between the two bears. Khutzy repeatedly stuck her head inside Skoki's mouth, removed it, and then both bears burst into laughter. Poulsen assumed it was just a unique play behavior they had invented, possibly as an expression of trust. However, a few years later, she observed the behavior again with a different bear.

This time, the "throat-diving" behavior occurred with an eight-month-old female American black bear cub, Miggy, who had been rescued and sent to the Detroit Zoo, where Poulsen was her foster mother. They ate, played, and even napped together.

It's nice to know that the sound of laughter can cheer up dogs the same way it can lift our own spirits.

During one play session, just as Poulsen burst into laughter at the little bear's antics, Miggy tried to stuff her snout into Poulsen's mouth. Catching Poulsen by surprise, Miggy partially succeeded and immediately fell onto the ground laughing. She quickly repeated this behavior, but Poulsen, wiser now and not wanting the cub's snout in her mouth, pulled back at the last second. Still, after each attempted throat-diving session, Miggy laughed as if it was the funniest thing in the world.

Poulsen had now observed the "throat-diving" behavior in three individuals (Khutzy, Skoki, and Miggy), two species (grizzlies and black bears), both sexes, and two different age groups. Although the sample size was still too small to draw any definitive conclusions, the three instances were enough to suggest to Pouslen that throat diving might be widespread play behavior among bears. When cubs are young, they sometimes smell their mother's breath, presumably to get a sense of what mama bear was eating, so Poulsen speculated that breath smelling might be the origin of throat diving. Regardless of the origin of the odd behavior, we'll probably never know why bears find it so funny. Perhaps sticking your head in your friend's mouth and not being eaten is simply hilarious (if you are a bear).

Humans may not "get" the joke, but to bears there's nothing funnier than a couple of friends mouthing each other's heads!

Laughter Is a Dog's Best Medicine

Many people believe that their dogs "laugh" when playing, and now science is proving them right. At the Animal Behavior Center in Spokane, Washington, researchers Patricia Simonet, Donna Versteeg, and Dan Storie discovered that dogs pant in a specific way when they want to initiate play. To the human ear, it sounds pretty much like a regular pant. But when scientists used a spectrograph to analyze various pants, they found that this particular pant had a different pattern of frequencies. When dogs hear it (named the "dog-laugh" by researchers), they respond favorably, with play-bows or play-chasing.

As part of their research project, Simonet and her colleagues recorded dog-laugh panting and played the recording to shelter dogs. The dogs responded by wagging their tails, playing, and engaging in other positive social behaviors. The recording also seemed to help decrease stressful behaviors. Researchers are hopeful that dog-laugh recordings will soon be widely used to soothe shelter dogs, which could potentially result in faster adoption times.

2. A GENEROUS NATURE

RECIPROCITY AND COOPERATION

GIFT-GIVING CROWS

Derek was a crow who rarely showed up without a gift, and Amanda was a wildlife rehabilitator who had done him a favor. Like other wildlife rehabilitators, Amanda cares for sick, injured, and orphaned wild animals with the goal of releasing them back into their natural habitats when they are ready. After being released, rehabilitated animals will sometimes linger in the neighborhood and return to visit the people who helped them. And every once in a while, these animals show up bearing gifts. This is what happened to Amanda after she helped Derek recover from a wing injury.

After receiving the injured crow, Amanda named him Derek. He quickly bonded with her entire family, including her black lab, Jake. At times when Amanda was away from home, Derek would show as much excitement and pleasure at her return as any dog would. He vocalized softly and rubbed his head against her hand, seemingly delighted to be back in her company. Derek also had a thing for Jake. He loved being near him and would gently peck at his paws and try to preen his fur. Luckily, Jake welcomed Derek's affection. In fact, they had a favorite game in which Derek held onto Jake's collar and rode him like a horse as Jake trotted around the yard.

Derek was released back into the wild once his wing healed. But, having developed a fondness for his

Crows are social animals that give one another gifts as a way to establish and maintain their bonds. Sometimes they even give gifts to other species, such as humans, cats, and dogs.

human family, he loitered in the local woods for a few months, staying close enough to call out to them and visit. Amanda often ate breakfast on her terrace, and one morning, when she ran inside to answer the phone, Derek swooped down from a nearby tree and stole the last bite of her croissant. Five minutes later, he reappeared and presented a leaf to her. During the next few weeks, Derek appeared every morning at breakfast time bearing gifts: leaves, twigs, acorns, colorful bits of plastic, even a key Amanda had lost a long time ago, and an occasional dead beetle. Amanda reciprocated by leaving him a bite of her breakfast. Sometimes Derek showed up in the afternoon when food wasn't present and simply left a gift on the table.

Derek's behavior isn't especially unusual. Stories both ancient and modern recount crows' propensity for giving gifts, especially in response to food. In 2015, one such story went viral on the Internet. Gabi Mann, an eight-year-old girl from Seattle, would sometimes accidently drop food while waiting for the school bus. When the crows nearby noticed, they descended to score a free snack. Gabi was charmed, and she started deliberately feeding them with bits of her school lunch.

Before long, Gabi and her family started feeding crows at home by filling an old birdbath with peanuts. Whenever the crows ate them all, one would return with a gift and leave it in the birdbath. Over time, these gifts included rusty metal, sea glass, old buttons, beads, and other tokens of appreciation. Once, a crow even returned with the missing lens cap from Gabi's mother's camera.

In other accounts, however, it seems like crows express gratitude for acts of kindness that have nothing to do with food. In John Marzluff and Tony Angell's book *Gifts of the Crow*, the authors describe how an ornithologist discovered and freed an injured crow who was stuck upside down in a fence. After a few minutes of recovery, the crow was able to fly away. But that was not the end of the story. From time to time the crow returned, leaving gifts at the exact spot on the fence where he had been rescued.

Crows aren't the only animals that give gifts. Scientists have observed similar behaviors between members of a variety of insect, spider, fish, bird, and mammal species. Most often, gifting occurs between mates or prospective mates. These are called "nuptial gifts," which range from food to inedible tokens, such as the pebbles that male Gentoo penguins

As part of their courting ritual, male Gentoo penguins present nest-building pebbles to females. Some males appear to be quite selective and spend time searching the beach for the perfect stones. Others take short cuts and steal their "engagement tokens" from other couples' nests!

present to females as part of their courtship ritual. Less often, animals will give gifts to other members of their group (friends, perhaps?), and occasionally, like the crows that befriended Gabi and Amanda, they will even present a gift to a member of another species. For example, there are numerous reports of dolphins presenting gifts of eel, squid, octopus, and fish to humans. And many people who share their homes with domestic cats tell stories about the dead rodents, birds, snakes, and other "gifts" left for them by their feline family member.

Reciprocity—the practice of exchanging resources or favors with others for mutual benefit—is an evolutionary adaption that increases an individual's chances for survival, especially in social species. When animals reciprocate positive actions, they strengthen the bonds of their relationships, just as humans do when lending a helping hand, exchanging gifts, or inviting friends to share a meal. It turns out that gift-giving isn't just a human cultural tradition, but something deeper and more prevalent: an ancient gesture of reciprocity that connects us with the larger fabric of nature.

a leopard seal's approval

Animals give "gifts" even in the remotest regions of Earth. In Antarctica, for example, a leopard seal presented unexpected gifts to a polar photographer.

Paul Nicklen was swimming with leopard seals while filming them, and many of his fellow scientists worried that the seals would attack Nicklen. That did not turn out to be the case. In fact, one large female leopard seal—with a head bigger than a grizzly bear's—seemed especially taken with Nicklen and began trying to feed him.

She brought him penguin after penguin—a favorite meal for leopard seals. At first, she brought him live penguins, which immediately swam away when Nicklen didn't respond like a leopard seal by eating them. Eventually, she changed her strategy: she caught a penguin, tired him out, and then presented him to Nicklen. When Nicklen again didn't respond, she decided to make it really easy—by bringing him dead penguins. Of course, Nicklen didn't accept those either. At this point the leopard seal backed off. She floated in the water for a few minutes, looking at Nicklen as if she couldn't believe he had refused her offerings. Finally, she grabbed a penguin and flipped him onto Nicklen's head, in an apparent attempt to force feed the photographer. That strategy failed, too.

Leopard seals—which can weigh up to thirteen hundred pounds and grow to more than ten feet long—can be intimidating. But one female leopard seal showed an arctic photographer her soft side—by trying to feed him penguins.

Although Nicklen didn't respond in the way the leopard seal seemed to want him to, he and the seal parted amicably. Nicklen now has the distinction of being the first human known to have received gifts from a leopard seal.

PULLING TOGETHER

So much of what humans have accomplished as a species has been the result of cooperation. Everything from astrophysics to zoology has depended on people working together and sharing discoveries and resources. Until recently, we thought of cooperation as one of the qualities that distinguishes humans from other animals. But many scientists now believe that cooperation and sharing evolved in other animals for the same reason they evolved in humans—to help ensure that the species is successful. It simply makes sense, from an evolutionary perspective, for members of certain species to share and cooperate if they want to survive and thrive.

Primatologist Frans de Waal and his colleagues conducted experiments to explore cooperation in capuchin monkeys, teaching them to reach cups of food on a tray by pulling on a spring-loaded bar attached to the tray. The tray was too heavy for just one monkey to move on its own, so there was an incentive to work together. During one of the experiments, two females, Bias and Sammy, worked together to pull the tray within reach. But Sammy, in a rush to get her food cup, released the bar and the tray bounced out of reach before Bias could get her cup. Bias screamed with apparent indignation while Sammy enjoyed her food. But Sammy soon gave in and helped Bias pull the tray back up so she could get her food, too.

Monkeys seem to believe that one good turn deserves another, and if it's time for you to return a favor, you better pay up!

Sammy didn't need to help Bias. There was nothing in it for her, as she had already retrieved her own food. When she stopped eating to help Bias, it likely meant one thing: she was responding to Bias's agitation. At least that was de Waal's hypothesis. He believes Sammy felt that she owed it to Bias to help retrieve her cup of food because Bias had already helped her get her own cup. This experiment suggests that some social species, like these capuchin monkeys, remember and appreciate favors the same way humans do, and they feel resentful when others don't pay back their favors. On the flip side, many species also appear to feel indebted when they are the recipients of favors, and return the favor when given a chance.

In a similar experiment, undertaken by de Waal, psychologist Joshua Plotnik, and other researchers, pairs of Asian elephants were taught to simultaneously pull both ends of a rope to move a platform containing food within their reach. If only one end of the rope was pulled, the rope slipped out and the platform wouldn't move. The elephants—each in their own separate fenced lane—quickly learned to coordinate their effort to pull the rope together, move the platform, and reach the food. They even showed an understanding that there wasn't any point to pulling until their partner was in position and ready to work together.

Even in the wild, the behaviors of both elephants and primates provide evidence that, like humans, some animals exchange favors based on long-term social ties rather than on what advantage they can gain in the moment. For example, elephants often demonstrate cooperative behavior, especially when it comes to helping out with parenting or tending to an individual in distress. Primates often cooperate too, even without any immediate reward in sight. A chimp will sometimes treat a friend to a long grooming session that isn't immediately reciprocated, but later the friend she groomed will be likely to remember her if she has food to share.

Elephants, known for their sensitive and empathetic natures, are also cooperative. In captivity, they have demonstrated a willingness to work together to solve a problem; in the wild, they have been seen cooperating with one another to rescue a herd member in distress.

For certain species of animals, including humans, generosity and cooperation seem to be more about remembering who is good to you over time—not just in the last five minutes—and being generally good to them in response, even if it means waiting before it's your turn to get your back scratched.

cooperating for the catch

In the coastal city of Laguna, in southern Brazil, people and bottlenose dolphins have engaged in cooperative fishing together for generations—at least for one hundred and twenty years, based on the historical evidence of a nineteenth-century letter that references the practice. The fishermen fish with nets, but due to the murkiness of the water, they can't determine the best locations to cast their nets. So, they partner with bottlenose dolphins. The fishermen wade into the water and wait for their partners to show up. Some days none show up; other days only uncooperative dolphins appear; but on good days, the helpful dolphins show up, ready to fish together. The fishermen insist that it's barely worth fishing if their friends aren't around to help.

Soon after arriving, the dolphins swim toward the fishermen, herding shoals of mullet into their nets. At the right moment, the dolphins signal the

Dolphins are not the only aquatic animals that have the capacity for cooperation. Scientist Simon Brandl studied cooperation in pairs of rabbitfish and found that while one forages and feeds, the other floats in an upright position watching for predators. This behavior, known as "reciprocal cooperation," is believed to require complex cognitive and social skills, capacities not usually attributed to fish. According to Brandl, this buddy-system strategy is advantageous because both partners appear to eat more when working together than when foraging alone.

fishermen—by performing a particular dive—that it is time to cast their nets. The fishermen claim that the direction and thrust of the dolphins' dives will tell them about the size of the school and the direction it is traveling in, which help them aim the nets.

Over the years, the local fishermen have given the dolphins names and even have favorite dolphins they prefer as fishing partners. As a result of their teamwork, the fishermen haul in nets full of fish. As for the dolphins, they get an easy meal out of it. Scientists believe that the net causes the fish to panic, which breaks up the school, leaving some fish on their own, and solo fish are much easier for dolphins to catch.

Dolphins cooperate in all sorts of situations, from assisting distressed pod members to partnering with people to catch fish.

3. Fair and Square

Playing by the Rules

Survival of the Fairest

Next time you feel irritated about getting the short end of the stick and wonder why it's so hard to let your indignation go, watch the YouTube video "Two Monkeys Were Paid Unequally." The popular video is an excerpt from primatologist Frans de Waal's "Moral Behavior in Animals" TED talk. With over ten million views, the video presents an experiment featuring two capuchin monkeys whose behaviors demonstrate that the drive for fairness is more universal than we thought, extending beyond the human species.

The idea for the now famous experiment arose from the observation of one of de Waal's students, Sarah Brosnan, while she was working with captive capuchins. She noticed that one of the monkeys would become agitated when he or she saw another monkey get a better reward for performing the same task. Known as "inequity aversion" in human economics, this behavior prompted Brosnan and de Waal to design an experiment to explore what appeared to be a sense of fairness in these diminutive primates.

When capuchins get the short end of the stick, they respond with indignation, just like people do.

The capuchins who participated were previously trained to give researchers small stones in exchange for food. During the experiment, the trained monkeys were paired and placed in adjacent cages so they could see each other. The TED talk video excerpt shows a few rounds of stone-for-food exchanges between the monkeys and a researcher. During the first round, the monkey on the left gives the researcher a stone and in return receives a piece of cucumber, which she readily eats. The monkey on the right then hands over her stone and is given a grape—a food that capuchins greatly prefer to cucumbers—and she happily eats her reward. The interesting thing is that the monkey on the left immediately notices—with a calm curiosity—that her counterpart has received a much better reward for the exact same trade.

In the second round, the monkey on the left is once again given a piece of cucumber in exchange for a stone. Even before the monkey on the right makes her exchange for a grape, the left monkey starts to throw a tantrum. She tastes the cucumber, then takes it out of her mouth and throws it at the researcher. She pounds the floor with her fists and rattles her cage. She repeats this same performance in the third round—clearly outraged that she is not getting a grape like the monkey on the right.

At this point, most people viewing the video burst into empathic laughter because the monkey's reaction is so incredibly humanlike. The monkey on the left is clearly getting shortchanged. She is performing the same work as the other monkey—exchanging a stone for food—but isn't getting the same reward. When she throws the cucumber at the researcher, viewers can't help but imagine her shouting, "You can take this job and . . ."

> The more effort it took to receive the reward, the more sensitive a monkey was to seeing another monkey get something better.

Brosnan and de Waal's experiment showed that monkeys' sensitivity to fairness was linked to effort. If they did not work (make an exchange) but were given unequal food portions, the monkeys didn't have negative reactions to any difference in the rations. However, if treats were used as payment for work—for the exchange of the stone—then the capuchins did notice inequities. In fact, the more effort it took to receive the reward, the more sensitive a monkey was to seeing another get something better. This drive to receive equal reward for equal effort is known as "first order" fairness.

Brosnan set up a similar experiment with chimpanzees, using grapes and carrots, with grapes once again being the preferred food reward. Like the capuchins, the chimps that received carrots rather than grapes for equal work were outraged and refused to continue with the experiment. But unlike the monkeys, the chimps that received the grapes were also upset, appearing to be uncomfort-

Primatologist Frans de Waal believes that fairness is an ancient capacity that probably arose out of the need to preserve harmony when confronted with resource competition.

able with the reward inequity. Bonobos (previously known as pygmy chimpanzees) have also reacted with similar uneasiness to unequal reward distribution, refusing a treat for themselves unless the other bonobos present received one as well. This kind of reaction—which transcends self-interest—is known as "second order" fairness, and has so far been observed only in humans and apes.

Corvids—the family that includes birds such as crows, ravens, and magpies—are often called "feathered apes" because of their amazing cognitive skills. They were put through a similar test by University of Vienna biologists Claudia A. F. Wascher and Thomas Bugnyar. Just as in the monkey and chimp experiments, researchers began by training the crows and ravens to exchange pebbles for food. They then tested same-species pairs in a series of exchanges in which sometimes both birds were rewarded, and sometimes only one was. The deprived birds reacted similarly to the monkeys and chimps: they were disgruntled by the inequity. After several inequitable exchanges, they stopped participating consistently in the experiment. While this was the first research to demonstrate that non-mammals react to inequity, the outcome isn't so surprising. After all, like monkeys and chimps, crows and ravens are cooperative species that form alliances and share resources, so a sense of fairness would benefit these birds in the same way it benefits other social animals.

Crows and ravens want to be treated fairly, and quickly get resentful when not rewarded equally for the same task.

Exploring still other species, Friederike Range, also from the University of Vienna, discovered that dogs have a sense of fairness, too. Range and her colleagues tested dogs that had already mastered the "paw" command, where a dog gives an "asker" his paw. The dogs tested in the experiment all had a history of freely offering their paw when asked, without any reward. But that behavior changed when one of these dogs was placed side-by-side with a dog

Our good-natured best friends don't seem to mind if others get a better reward for doing the same work. As long as they all get rewarded, they're happy.

that was rewarded with food. At first, the unrewarded dogs merely hesitated to shake, but eventually they started showing signs of frustration, and then they stopped cooperating completely.

Range's study demonstrates that dogs, like primates and corvids, care about fairness, although it turns out that they are not quite as picky about the equity of the reward. Range did a second paw-command experiment in which she rewarded one dog with bread and the other with sausage. Unlike the monkeys, most dogs kept working even if they didn't get the better reward.

Researchers are starting to believe that a sense of fairness is most likely an evolutionary adaptation that developed in certain social species. When discussing the capuchin fairness experiments, for example, de Waal compared the similarity between the responses of the monkeys and chimps to the protesters in the Occupy Wall Street movement: all were objecting to the idea of economic injustice. As de Waal pointed out, if eating were the only thing that mattered, the monkeys would have as readily accepted the cucumber as the grape. But fairness matters, too. Over time, if inequities are allowed to perpetuate they become exploitative and have a detrimental impact on the health of individuals, groups, and perhaps even entire species.

So, getting miffed over inequities may be a survival strategy. Perhaps evolution has favored emotions like indignation because it invokes behaviors that foster fairness. And when a species behaves fairly, not only is each individual better off, but the species as a whole is, too.

De Waal compared the similarity between the responses of the monkeys and chimps to the protesters in the Occupy Wall Street movement: all were objecting to the idea of economic injustice.

FIDO'S Fair Play

Dogs don't just want to be fairly rewarded when working; they also want to be fairly treated when playing. During play, canids (the family of animals that includes dogs, wolves, coyotes, foxes, and others) often engage in mock biting, mounting, body slamming, and other rough-and-tumble actions that might cause harm if not subject to some kind of behavioral parameters. After years of observing canids, ethologist Marc Bekoff believes they do indeed have rules that govern what constitutes fair play.

Canids start by communicating clearly (through gestures) that they want to play. These gestures are different from those used to provoke a fight. Their "rule book" for play includes a series of agreements that keeps play safe and peaceful. As examples, they have self-handicapping and role-reversal strategies to make the playing field level between dogs of different sizes, strengths, ranks, and experience. If they get too rough, they "apologize" by bowing, and seem to forgive one another if mistakes are acknowledged. In canid "culture," those who don't play fairly are often ostracized from the pack, whereas those who play fairly appear to develop deeper bonds of trust with other pack members. This results in stronger, healthier, and more cohesive packs.

Dogs, wolves, and foxes—among other canids—like to play, but they like to play fair. They follow unspoken rules of behavior that distinguish playfulness from aggression.

Crimes and Misdemeanors

Just as fairness isn't limited to humans, neither is cheating. There are plenty of animals that are known to cheat. Crows, for example, often steal food from one another, and some even work together to pull off a heist. A favorite corvid teamwork technique involves one crow pulling the tail of an animal who is eating. When the unsuspecting victim turns around to see who tugged his tail, the other crow dives down and steals his food.

Hanging out with a cheater might be okay if you're a cheater yourself. But cognitive biologist Jorg Massen discovered that if you are trustworthy, you don't want to be around cheaters. At least, this is what he found to be true for ravens. In an experiment that looked at cooperation in ravens, Massen tested whether pairs of ravens could work together to achieve a mutually beneficial outcome. As in the cooperation studies presented in chapter two, the ravens had to simultaneously pull both ends of a rope to slide a platform with two pieces of cheese into reach. If they didn't work together—if only one bird pulled the rope—the rope would slip off the platform, leaving the cheese out of reach for both.

The ravens quickly figured out how to work together. Given how smart ravens are, that wasn't so surprising. What did surprise Massen was what happened when some of the birds cheated and took both pieces of cheese. The ravens who were robbed became reluctant to work with their cheating partners. Although refusing to work with the cheater meant that neither bird got a cheese reward, the raven who was robbed at least got to ensure that the cheater was punished.

Ravens would rather go hungry than work with a cheater.

When it comes to cheating, monkeys may have a double standard: they insist on equal pay for equal work, but given a chance to cheat the system, they will. Behavioral economist Keith Chen and psychologist Laurie Santos found this out when they trained capuchin monkeys to use silver discs as currency as part of a study examining the potential origins

It appears that humans aren't the only animals to practice the oldest profession in the world. Scientists have observed chimpanzees, Adélie penguins, and capuchin monkeys exchanging items of value—such as food, nest pebbles, or "Monkey Market" tokens—for sex.

> The monkeys weren't always law-abiding, jumping at opportunities to steal tokens from the researchers and one another, and even trying to get away with "counterfeiting"!

of economic behavior. After teaching the monkeys about the value of the tokens and how to use them, the researchers created a Monkey Market where the monkeys could go to purchase grapes, apples, cucumbers, and other treats, all displayed neatly on trays. Each monkey was given a wallet with twelve silver tokens.

Sometimes, like any honest shopper, the monkeys behaved honorably, surveying the offerings, carefully examining and sniffing the goods, and then making purchases. But other times the monkeys weren't so law-abiding. They would jump at opportunities to steal tokens from the researchers and one another. The researchers humorously described one capuchin as trying to get away with "counterfeiting," by trying to pass off sliced cucumbers as tokens. On one of the most memorable days at the Monkey Market, a capuchin flipped over a tray of tokens, and as they scattered he and others stole as many tokens as they could. The researchers tried to get the stolen tokens back, but the monkey thieves refused to hand them over. The researchers resorted to bribing the monkeys with food in order to retrieve the tokens (who said crime doesn't pay?). But here's the real kicker: during the chaos of the token heist, Chen witnessed another crime occurring simultaneously—a capuchin was trying to exchange a token for sex! Clearly, money changes everything.

Researchers were surprised by how quickly capuchin monkeys were able to grasp the concept of money and devise cunning ways to accumulate it and use it to their own advantage. Credit: Laurie Santos

4. Stand by Me

Friendship

They Get by with a Little Help from Their Friends

A primatologist once joked that friendship was the "F-word," a word that scientists were reluctant to use in public when describing supportive, loyal, affectionate relationships between two unrelated, unmated animals. Calling such relationships "friendships" was considered anthropomorphic—projecting human characteristics onto animals. However, the more researchers learn about animals, the more it seems that the word *friendship* perfectly describes the close relationships that some animals share. Many animals intentionally seek out other unrelated members of their species to groom, share food with, play with, and sleep with. Some even help each other and "have each other's backs" in conflict situations. These behaviors meet the criteria that we use to define human friendship, so why not use the same word to describe these social bonds between animals? Some scientists are finally ready to do just that.

Researchers have observed friendship in a broad range of animals, including primates, birds, elephants, whales, dolphins, horses, cows, and other species. Even rattlesnakes may have friends. Melissa Amarello, a researcher who studies the social lives of snakes, discovered that some rattlesnakes associate with other rattlesnakes in close proximity for what appears to be nothing more than the pleasure of company. They spend time together, basking in the sun, and sometimes touch in non-aggressive ways. Some female rattlesnakes even babysit one another's offspring.

One of the hallmarks of friendship is loyalty, which has long been recognized in geese. During migration, if a member of the flock is injured or becomes ill, one or more other geese leave the flock and follow the incapacitated bird to the ground, where they stay with it until it either recovers or dies.

Compassion is another aspect of friendship. Scientists have discovered that female elephants help care

They might not be warm and fuzzy, but rattlesnakes have more in common with us than previously thought. They have homes and mates and care for their young; they have favorite places to eat and hang out; and some even have friends that they visit throughout their lives.

for one another's calves, tend to the old and injured, and develop close relationships. By studying how Asian elephants respond to herd members in distress, behavioral ecologist Joshua Plotnick and primatologist Frans de Waal learned that elephants behave compassionately. Over the course of a year, they studied twenty-six adult female and juvenile elephants in Thailand. They watched for signs of distress, which include certain vocalizations, flared ears, erect tails, and other behaviors. When an elephant exhibited these distress signs, the researchers noticed that nearby herd members approached the agitated elephant, caressed her, and vocalized softly. The scientists were impressed by how consistently the elephants consoled one another when upset. They rarely saw a distressed elephant neglected by her herdmates.

Sharing food is a universal gesture of friendship—even among elephants.

Sadly, one of the most stressful experiences for elephants is captivity, where they are often abused, isolated, neglected, and consequently depressed. An especially touching story about how elephant friendship in captivity can be a source of mutual compassion involves Shirley and Jenny, two former circus elephants. After a twenty-two-year separation, the elephants were reunited at a sanctuary and recognized each other immediately. They trumpeted and rumbled to each other, and excitedly reached out to touch each other with their trunks, proving that when it comes to friends, it's true that elephants never forget. The two old friends so passionately wanted to be together that they bent the steel bars on the barrier that initially separated them while the caregivers made arrangements for a successful reunion. As soon as the gate was opened, the two elephants became inseparable, and they never left each other's sides.

A group of female humpback whales met up once a year to share a few meals, swim, and sing together, doing so for six years in a row.

Elephants aren't the only animals with best friends. Krista McLennan, an animal welfare researcher, observed that cows tend to form close relationships with other members of their herd. In fact, cows seem to have preferred pals that they consistently spend time with. When McLennan did a study that measured the heart rates and cortisol levels of cows when they were with their best friends and when they were separated, she noticed that their heart rates and stress levels were noticeably reduced when they were with their buddies. Clearly, if you are a cow, being with friends is good for your health.

Just like us, when cows are anxious there's nothing like the reassuring and calming presence of a good friend.

While it's not possible to measure heart rates and stress levels of humpback whales, they seem to have close friendships, too. Researcher Christian Ramp discovered that a group of mature female humpback whales, all roughly the same age, met up once a year in the Gulf of St. Lawrence to share a few meals, swim, and sing together. They were observed doing this for six years in a row. Ramp doesn't know how long these reunions had been going on before his observations, or how the whales find one another for these gal pal reunions. Some scientists believe that the whales' songs might help them reunite. As to what draws these whales together every year, current whale brain research may help clarify the matter.

In 2006, two scientists, Patrick Hof and Estel Van Der Gucht, discovered spindle cells in the brains of humpback and other whales. In humans, these cells are associated with emotion, empathy, and social interaction, and have often been described as the "cell that makes us human." It looks like whales may have three times as many spindle cells as humans do (even after adjusting for brain size), and they may also have had the cells for twice as long as humans, evolutionarily speaking. So perhaps these yearly reunions happened for the same reason that human girlfriends get together: they simply enjoy one another's company. At the very least, scientists are beginning to believe that whales are capable of the kind of emotions and social attachments that develop into friendships, and possibly even love.

Humpback whale brains may have three times the number of spindle cells—a specific kind of brain cell associated with emotion, empathy, and social interaction—as humans. This suggests that their brains are wired for friendship and possibly even love.

THE PEACEABLE KINGDOM

About ten years ago—thanks to the viral nature of a good story on the Internet—the world learned that animal friendships are not restricted to members of the same species. Nature photographer Norbert Rosing was responsible for one of the first inter-species friendship stories that spread over the web. Rosing was in Churchill, Manitoba, where he saw a large polar bear approach a group of tethered sled dogs. Occasionally polar bears will kill a dog, so Rosing braced himself for the worst. All the sled dogs started barking and behaving aggressively, except for one. Much to Rosing's surprise, this one dog acted friendly and started wagging its tail, as if he were reuniting with an old friend. The bear took notice and decided to ignore the other threatening dogs, walking instead toward the welcoming dog. Rosing held his breath as the bear approached.

What he saw next surprised him even more than the dog's demeanor. Instead of becoming the bear's lunch, the dog became the bear's buddy, and the two animals began to tumble around in the snow, playing like puppies and cubs. While they were playing, the bear even rolled over on his back, which is a submissive posture for both dogs and bears. Rosing couldn't believe what he was seeing. He documented the encounter through a series of photos that would later become famous. After about twenty minutes, the bear left, only to return to play with his new canine pal the next day and for several days afterward.

Another surprising story of apparent friendship between two species happened off the coast of Hawaii. Humpback whales and bottlenose dolphins were observed engaging in a game scientists described as "lift-and-slide." The whales lifted the dolphins out of the water on their rostrums (the beaklike part of a whale's head) and the dolphins slid down the backs of the whales into the water. It looked like both animals were having fun, but scientists had to consider two other possible explanations: aggression or altruism. The posture of the whales and cooperation of the dolphins ruled out aggression; and the dolphins seemed healthy, which ruled out the possibility that the whales were helping injured dolphins. It seemed the only plausible explanation was interspecies play, possibly arising out of friendship.

Even so, these encounters between the sled dog and polar bear and the whales and dolphins appear more like impromptu interspecies play rather than true friendship. It is most often in captivity that we observe animals of different species forming long-lasting bonds with one another. In her book *Unlikely Friendships*, author Jennifer Holland presents stories about such relationships, including one about a black cat that wandered into a black bear's enclosure at the Berlin Zoo. The two animals—with only the color of their coats in common—quickly became close friends, sharing meals and even sleeping together.

In another story, Holland describes a relationship between three male predators—a bear named Baloo, a lion named Leo, and a tiger named Shere Khan—known as the "BLT" trio. All three animals had been kept in deplorable conditions in the basement of a private home and were discovered during a police drug raid. They arrived at a sanctuary as young animals and were housed together. The BLT trio grew up as inseparable best friends, playing together, napping side by side, and grooming one another.

After these three animals were rescued from a drug dealer's basement, they lived together at a sanctuary for fifteen years. Sadly, Leo passed away in 2016. Despite the deep bond they shared with their lion "brother," Baloo the bear and Shere Khan the tiger are doing well. Credit: Noah's Ark Animal Sanctuary

Although these playful encounters and friendships are unusual, they all involve mammals, which share many behavioral traits across species. When friendships develop between species of different classes, the relationships are even more astonishing. For example, Holland recounts the story of Sobe and Jo, two animals rescued by an animal welfare advocate. Sobe, an iguana, moves freely around his caregiver's home, while Jo is the especially affectionate house cat. The two animals—a reptile and mammal—appear to have genuine affection for each other: Sobe lets Jo rub against him, groom him, and even play with his tail. And at nap time, Jo is often found curled up next to his scaly friend, snoozing and purring.

As odd as Sobe and Jo's relationship is, the reptilian-feline duo still shares one substantial commonality: both are terrestrial animals. In contrast, one of the most unusual friendships in Holland's collection is between Chino, a golden retriever, and Falstaff, a koi fish—a large carp commonly found in outdoor water gardens. Whenever Chino visited Falstaff, the fish would immediately swim over to his furry friend and the two would greet by touching noses. During his visits, Chino would lie at the edge of the pond, peacefully watching his aquatic friend for up to half an hour. Sometimes Falstaff would gently nibble on Chino's paws. According to Chino's human family members, visiting Falstaff was one of the dog's favorite activities (an opinion they formed based on his wagging tail).

Humans have many interspecies friendships, but these are usually with domesticated species. Every once in a while, however, a wild animal and a human become friends. Such a friendship began in 2011, when a seventy-one-year-old retired Brazilian bricklayer named João Pereira de Souza discovered an oil-soaked Magellanic penguin lying on the beach. João cleaned up the sick bird, fed him sardines, and returned the penguin to the beach, expecting him to swim back home. But the bird had found a friend in João and refused to leave. So João named his new buddy Dindim.

On rare occasions, humans and wild animals form friendships. One injured penguin found both a rescuer and a lasting friend in a retired Brazilian bricklayer.

Eventually Dindim swam away, but then he returned, beginning a visitation cycle that has gone on for years. When Dindim leaves, he is gone for weeks or even months, but he always returns to visit his friend, João. Upon every arrival, the little tuxedoed bird vocalizes excitedly and gently nibbles his human pal, which is a preening behavior usually reserved only for a penguin's intimates. After each of their emotional reunions, João and Dindim spend time doing what friends do: they take walks on the beach, swim together, and just hang out.

If a cat and an iguana can nuzzle each other and nap together, and a dog and fish can "kiss" upon meeting at the boundary between their terrestrial and aquatic worlds, then it's time for humans to take a lesson from other animals in how to get along.

Some unlikely friendships amaze us because they call to mind the biblical "peaceable kingdom" in which predators and prey live in harmony. Others demonstrate that very different species of animals may have more in common than previously thought, and suggest that friendship can reach across wider divides than we ever imagined. Finally, all of these interspecies friendships ask us to think about the differences that sometimes keep humans apart. If a cat and an iguana can nuzzle each other and nap together, and a dog and fish can "kiss" upon meeting at the boundary between their terrestrial and aquatic worlds, then it's time for humans to take a lesson from other animals in how to get along.

Interspecies friendship raises intriguing questions about animal communication, such as: how do different species learn to read one another's signals? It seems that when raised together in captivity, animals sometimes develop a common language that they use to establish the rules of their interactions, such as how to play together, when to give give each other space, and when it's okay to snuggle.

5. FOR THE FUN OF IT

PLAY AND IMAGINATION

ALL WORK AND NO PLAY ISN'T NATURE'S WAY

Once, when paraglider Tim Hall was paragliding above a three-thousand-foot cliff, he encountered a group of ravens who were diving and doing barrel rolls. Soon another raven—carrying a twenty-foot-long plastic streamer in his beak—joined the group. This raven plunged through the air with his streamer trailing behind and then passed the streamer to another raven, who did the same thing. For about twenty minutes, the ravens took turns passing the streamer while soaring above the cliff.

> Ravens have been observed doing barrel rolls, windsurfing, and playing tag for fun.

In *Gifts of the Crow*, John M. Marzluff and Tony Angell describe another example of ravens using an object in an unexpected way. Eight ravens were observed gripping curved pieces of tree bark in their talons and using them to windsurf strong gales in the Rocky Mountains. They spread their wings, launched into the air, and used their feet to adjust the angle of their "surfboards" to soar, dive, and slide along the wind currents. Ravens have yet another in-flight game, known as "stick tag," in which one raven holds a stick while being chased by others who try to steal it. At some point, the stick-holder drops the stick and another raven catches it midair, becoming the pursued.

Why do ravens engage in these activities? Bernd Heinrich and other scientists believe that they are simply having fun. But ravens aren't the only animals who know how to play and enjoy themselves. A wide variety of species have been known to engage in versions of chase, tug-of-war, peek-a-boo, hide-and-seek, keep-away, and king-of-the-mountain. Wrestling, tickling, and teasing games are also played by many animals. From predictably playful bear cubs—who resemble toddlers when sliding down snowy hills and engaging in rough-and-tumble games—to play-fighting female paper wasps rehearsing behavior they'll need as future queens, the more scientists look for play across the animal kingdom, the more they find it.

For a long time, people believed that play was limited to humans and other mammals, but current research has revealed that birds, reptiles, fish, and even invertebrates—such as octopuses, spiders, and the previously mentioned paper wasp—also play. In his book *What a Fish Knows*, animal behaviorist Jonathan Balcombe describes captive fish riding bubbles, playing with an aquarium thermometer, and even engaging in an ambush game with a cat who liked to steal a drink of water from their tank. The fish would hide until the cat approached the top of the tank and then dart up to the surface to startle the thirsty feline. Both the fish and cat repeated the exchange over and over, and neither was hurt—two conditions that suggest it was play rather than territorial aggression.

According to researcher Vladimir Dinets, alligators and crocodiles play, too. Dinets conducted a comprehensive study of wild and captive crocodilians and found that they slide down slopes, surf waves, chase one another, and play with toys for fun (even appearing to prefer pink toys). Crocodilian object play is now so widely recognized that many zoos provide toys as part of habitat enrichment.

Finding play behavior in so many different kinds of species inspired scientists to investigate the purpose of play. Years of research have shown that play functions to strengthen social bonds, provide opportunities to learn right from wrong (such as how hard to bite and what gestures are off limits), increase physical fitness, and acquire and practice important skills in a safe arena. In general, animals play under the same conditions as humans—when "work" is done, meaning after other needs such as safety, food, and shelter are met and when everyone

Recent research suggests that fish are not only sentient but also hunt cooperatively, use tools, have social bonds, and even play.

Bear cubs are especially playful, often resembling human toddlers when they play fight.

is relaxed and feeling good. And research suggests that they play for the same reason as humans: because it's fun.

In another book, *Pleasurable Kingdom: Animals and the Nature of Feeling Good*, Balcombe presents several reasons why he believes animals have fun when they play. First, most animals simply look like they're having fun. Second, animal play is similar to human play, and since humans enjoy playing, it is reasonable to assume that other animals enjoy playing, too. Third, in captivity, animals often choose playing over eating unless they are very hungry. Sometimes captive animals will even work for a chance to play. Balcombe describes a captive orangutan at the Saint Louis Zoo who would clean his enclosure in exchange for time playing with his whistle. Fourth, there are chemical changes—such as an increase in production of the feel-good neurotransmitter dopamine—in the brains of animals who are playing or anticipating an opportunity to play.

During play, juveniles of many species—from ravens to lions—often stalk one another to try to steal a coveted stick or other object.

Based on these observations and the biochemical evidence, it is very likely that playing is pleasurable for animals. And given all the physical and social skills learned from and improved by play, scientists believe that play increases an animal's chances of survival. Evolution, therefore, favors those who play more and have more fun. (Mother Nature must have been in an especially good mood when she designed a system that rewards animals for having fun.) So, the next time you see a couple of chipmunks chasing each other around a tree or your cat stalking her own tail, ask yourself when you last played. If it's been a while, take a lesson from the animals and do something just for the fun of it.

Snow Days

Just like people, animals can be inspired to play by all sorts of situations and circumstances, including winter weather. In fact, snow seems to be irresistible to some. When it snows, prairie dogs leave their dens and run and tumble through the snow. According to scientist Con Slobodchikoff, the prairie dogs purposefully fall onto soft drifts and slide down little snow-covered hills on their bottoms. Japanese macaques—the northernmost nonhuman primate—get creative with snow in another humanlike way: they make and play with snowballs. They collect the snow, roll it into spheres, and then roll them down hills, carry them around, or even sit on them. As far as we know, they have yet to discover the projectile potential of snowballs, but perhaps they just need a little more time.

In Alaska and Canada, wild ravens are frequently seen sliding down snow-covered roofs on their

Japanese macaques have mastered the art of making snowballs, but they have yet to discover the fun of a snowball fight.

breast and belly feathers. Like children sledding on hills, when the ravens reach the bottom of the roof, they climb or fly back to the top to do it all over again. Ravens in Bernd Heinrich's aviary in Maine developed their own "sledding" style in which they hold sticks in their talons while sliding down mounds of snow on their backs. No one knows why these Maine ravens carry sticks as they slide, though Heinrich suspects it might be a kind of social display to "show off" and draw attention to themselves.

In 2012, one Russian crow became an Internet sensation after being videotaped while "snowboarding" on a roof. Aleksey Vnukov uploaded a video of a crow using a jar lid as a makeshift snowboard. The crow carefully positioned the lid at the peak of the roof, stood on it, slid to the bottom edge of the roof, and then climbed back up to take another ride. The crow looked so much like he was snowboarding that the video—watched over 1.5 million times—soon earned the title "crowboarding."

In 2012, one Russian crow became an Internet sensation after being videotaped while "snowboarding" on a roof.

Other animals, such as penguins, polar bears, and otters, also enjoy sliding on snow, but snow isn't the only cold weather phenomenon that inspires animal play. In his book *Winterdance: The Fine Madness of Running the Iditarod*, author Gary Paulsen recounts a story about a musher who saw buffalo playing on ice. For over an hour the musher watched a small group of buffalo take turns building up speed by running along the shoreline of a frozen lake and then sliding onto the ice. As each buffalo hit the edge of the ice, he spread his feet apart, raised his tail, and bellowed as he slid and spun across its slick surface. Before he made it back to solid ground, another buffalo would slide onto the ice. The buffalo repeated their sliding game again and again. From the musher's perspec-

There's something about sliding across the frozen surface of a lake that is irresistible—perhaps even to buffalo.

tive, it looked like the buffalo were trying to see who could slide the farthest. While this interpretation may be a bit of a stretch, the most convincing explanation for their behavior is that they were simply having fun.

IT'S ALL FUN AND GAMES FOR THE DOLPHINS

Scientists have learned that many animals play, but there are a few species that appear to excel at the art of having fun. Dolphins, in particular, seem to have devised all kinds of ways to amuse themselves. Captive dolphins (as well as belugas) blow bubble rings and then play with them by batting them with their fins and swimming through the larger ones as if they were hoops. In the wild, dolphins leap and frolic in the surf and in the pressure waves created by boats and whales. Under the surface, they play tug-of-war and keep-away with strands of seaweed, and, as mentioned in chapter two, they even play catch with other species.

Another species that dolphins "play" with is the pufferfish, possibly to "get high!" The BBC used hidden cameras to capture footage of a group of dolphins passing around a pufferfish that they took turns gently mouthing. According to zoologist Rob Pilley, the producer of the BCC show, after mouthing the pufferfish, the dolphins just floated around near the surface of the water, seemingly fascinated by their own reflections. Pufferfish produce a toxic defensive chemical when threatened, but in small doses it seems to induce a trancelike state that dolphins appear to purposefully induce. Pass the pufferfish?

HANGING FOUR FOR FUN

Goats are known for their remarkable balance: they can climb sheer cliffs, rugged mountains, vertical dams, and even trees. California surfer Dana McGregor has now proven that they can also balance on a surfboard. In 2011, McGregor bought a female goat and named her Goatee. The goat accompanied him everywhere, including his favorite surfing spot, Pismo Beach. One day while hanging out on the beach with Goatee, he decided to teach her to balance on his board. Once she got used to the water (goats usually dislike water), McGregor taught her to surf. Because she wasn't able to paddle, McGregor would stand with her on the back of the board and paddle out past the breakers, waiting for a good wave. Then he and Goatee would ride the wave back to shore together.

When Goatee gave birth to a son, Pismo, McGregor introduced him to surfing while he was still young.

Goats have an exceptional sense of balance, so when California surfer Dana McGregor decided to take his goats surfing, they quickly got the hang of it. Credit: © Jonah and Lindsay Long (jonahandlindsay.com, with thanks to Dana McGregor, surfinggoats.com)

The early exposure gave Pismo the edge he needed to become an even better surfer than his mother. With McGregor on the board, little Pismo would spread his legs to steady himself as a wave approached. Then, they would ride the wave together. Once, just before they caught an impressive nine-foot wave, Pismo head-butted McGregor off the board and rode the wave to shore by himself. Pismo then looked back at McGregor—still in the water—with an expression of cocky triumph.

While teaching his goats to surf, McGregor never enticed them to learn with food as a reward, though he often hugged and praised them when they performed well. McGregor believes that Pismo enjoys surfing for the same reason he does: it's fun. It's also possible that the goats enjoy surfing because they do it well. Like humans, goats and other animals appear to experience what is known as *funktionslust*, a German word meaning "the pleasure taken in what one does best." Jonathan Balcombe points out that most people prefer playing games they are good at because performing well feels good, so it's likely that animals sometimes enjoy playing for the same reason. Goats excel at balancing, so McGregor may have picked the perfect "sport" for them.

BY ANY STRETCH OF AN APE'S IMAGINATION

Like people, animals engage in a wide variety of play activities, but what about imaginary play? One of the hallmarks of human play, especially in childhood, is the creative use of imagination to invent things, characters, and situations, and to attribute to them qualities and conditions that don't exist. A child might assert that a dragon lives in her closet, that a stuffed bear talks, or that tea fills an empty cup. For a long time, this kind of imaginary play was considered uniquely human, but now it looks like the great apes share the ability to engage in make-believe.

In the wild, scientists have observed juvenile chimps ascribing living characteristics to inanimate objects—pretending that sticks and small logs are infants. Researchers Sonya M. Kahlenberg and Richard W. Wrangham conducted a study on chimpanzee behavior in Uganda and documented numerous instances of chimps treating sticks as dolls. The chimps cuddle, cradle, and put their stick babies to bed in nests they make for them. Some chimpanzees even carried their sticks into their nests to sleep with them, reminiscent of the way children sleep with a favorite stuffed toy. The

Chimpanzees and other apes appear to have rich imaginations that might inspire their daydreams and play behaviors, and possibly even improve their problem-solving abilities.

researchers also observed the chimpanzees playing a version of the "airplane game" with their sticks—lying on their backs with their stick balanced across their upraised hands. Mother chimps play this same way with their babies.

Captive apes have shown that they engage in imaginary play, too. Viki, a chimpanzee raised by Keith and Catherine Hayes as part of a language-learning experiment, moved and gestured as if she were dragging an imaginary pull-toy around the house. Sometimes Viki appeared to pretend that the toy was stuck on something and would mimic actions, such as tugging on the invisible string, until she had freed it. On another occasion, Viki went even further, inventing an entire fantasy scenario. Again acting as if her pretend toy was stuck, she sat down and—placing one fist on top of the other on the imaginary string—leaned backward as if she were pulling with all her strength. Feigning defeat, Viki then called out for Catherine, who came over and mimicked untangling the string and freeing the toy. Viki then got up and happily resumed pulling her imaginary toy around the room.

Other captive chimpanzees, as well as bonobos, gorillas, and orangutans, do the same. Some pretend to eat food depicted in photos or play-act biting and being bitten by plastic animals or photographs of animals. Apes have also kissed, hugged, stroked, tickled, bathed, fed, and vocalized to dolls and stuffed toys; pretended to be "monsters" or other animals and characters while wearing masks; hidden invisible objects; and given imaginary objects as "gifts" to others.

Learning that apes can engage in this kind of imaginative play was startling to many scientists. After all, scientists and philosophers have long believed that one of the major features that sets humans apart from other animals is our imaginative capacity. Now that we know apes have an imaginative inner life, we have one more reason to do everything we can to prevent their extinction. This way we can continue to get to know and share our world with the only other animals (as far we know right now) that plays make-believe.

6. RANDOM ACTS OF KINDNESS

EMPATHY AND ALTRUISM

COLD NOSES, WARM HEARTS

If you live with a dog, try yawning in front of him and see what he does. Chances are your dog will yawn in response. We all know from experience that humans find yawning contagious, but it turns out that dogs, wolves, and other animals seem to be equally susceptible to catching a yawn.

Scientists call mimicry such as contagious yawning and similar mirrored behaviors or moods "emotional contagion." It's an empathetic sensitivity to recognizing, feeling, and sharing the emotions and sensations of others. In its simplest form, empathy is a physical response to another's movement or mood, such as your dog yawning after you do; in its more complex form, empathy is an example of "perspective taking," in which you imagine what another person might be feeling or experiencing. We have long known that empathy and perspective taking are human traits, but now scientists are discovering them in animals, too.

There are countless stories about dogs comforting their owners during times of grief and depression, and getting a "contact high" when playing with happy family members. Most dog owners have little doubt that their canine companions are empathetic. But such claims are often dismissed by scientists as

Most dog owners believe their canine companions can sense their moods, and some have even suggested that their dogs are so tuned into them that they are able to "read their minds." Science has yet to determine whether dogs are just particularly adept at noticing human patterns and anticipating behavior or if they have a kind of canine "sixth sense." After all, dogs can detect cancer, sense low blood sugar, and predict seizures, so it's not that far-fetched to imagine they can sense our thoughts, too.

anthropomorphism—attributing human traits to nonhumans.

To test this widespread belief in the empathy of dogs, researchers Deborah Custance and Jennifer Mayer studied the behavior of eighteen different dogs. They set up controlled sessions involving a dog, her owner, and a person unfamiliar to the dog (the "stranger"). The owner and stranger sat and talked and took turns displaying two contrived behaviors: a benign behavior, humming; and a distressing one, crying. The researchers wanted to see how each dog responded to the distressed behaviors, and if there was a difference in the dog's response based on whether his owner or the stranger displayed the behavior. The researchers believed that if the dog responded only to his owner's distress, the dog's response might be more about his own feelings than empathy. This belief was based on the premise that the dogs might become anxious themselves when their owners pretended to cry, so responding only to their owner could reflect nothing more than the dog's need for personal reassurance. But if the dog responded to both his owner and the stranger when each pretended to be distressed, then it was more likely to be empathy.

The majority of the dogs passed the empathy test, pretty much with flying colors. Each not only responded much more to the displays of distress than to the benign behaviors, but each also reacted to both the owner and the stranger. When either pretended to cry, the dog sniffed, nuzzled, and licked that person—behaviors consistent with an expression of empathy. Although Custance and Mayer don't claim that their study definitively proves canine empathy, they do believe that it's a step toward validating what dog lovers have long believed. As for the rest of us, we will continue to feel comforted when our dogs lick and nuzzle us when we're feeling down, even without definitive empirical evidence.

Cats Don't Want You to Know This

Dogs and people have been hanging out together for over thirty thousand years. But cats were domesticated only about nine thousand years ago, so it's not a surprise that cats and people are still a little bit of a mystery to each other. Beyond knowing what food

It turns out that cats really do care about what their human companions think.

and toys their cats like, and where they prefer to be scratched, most cat owners don't have a clue as to what goes on in the feline mind. Until recently, scientists believed that the mystery was mutual—that cats didn't know or care much about what was going on in the human mind either. But new research suggests otherwise.

Scientist Isabella Merola recruited thirty-six pairs of cats and their owners for a study on "social referencing," which is akin to empathy in that it involves assessing another person's (or animal's) emotional cues to obtain information about a situation. Infants, children, and adults do this, as do dogs and primates. But it was a surprise to learn that cats—with their independent and sometimes aloof demeanors—do so, too. Merola set up tests to see how cats reacted to their owner's feigned fear or delight in response to ribbons attached to a running house fan.

The owners either acted fearful and moved away from the fan with its streaming ribbons or they mimicked delight and moved toward it. Seventy-nine percent of the cats responded in a way that suggested they were attuned to their owner's reactions. The cats continually looked back and forth from the fan to their owners as if searching for emotional clues as to how to react. To some extent, they also modified their behavior accordingly. Both reactions suggest that cats do indeed care about what their human companions think and feel, and they even look to them for direction in unfamiliar situations.

In another study, researchers Moriah Galvan and Jennifer Vonk observed twelve cats to see whether they would change their behavior based on their owners' differing moods and emotions—and they did. When their owners were smiling or happy, the cats behaved affectionately, purring and rubbing against them, even climbing onto their laps. But if the owners frowned or spoke aggressively, the cats behaved less affectionately.

Although neither test is definitive, both suggest that cats may have a capacity for empathy—at least at the emotional contagion level. So much for the myth of feline aloofness.

The Compassionate Rat

Is empathy an evolutionary advantage? Charles Darwin thought so, saying that animals whose communities "included the greatest number of the most sympathetic members would flourish best, and rear the greatest number of offspring." Perhaps this explains why rats are such a successful species.

To test whether the prolific rat has an empathic nature, neurobiologist Peggy Mason placed pairs of rats in a clear acrylic box, confining one to a clear plastic tube with a door at one end that could only be opened from the outside while allowing the other rat to roam freely. Rats tend to avoid open spaces and prefer to stay in corners or travel along walls, yet in every test, the free-roaming rat left the safety of her corner and figured out how to open the door and release the trapped rat. After releasing the trapped rat, the two rats would often nuzzle each other.

Rats are unexpectedly kind and empathetic—they often value helping other rats more than receiving food rewards for themselves.

According to Mason, the free rat appeared to recognize the trapped rat's distress, empathized, became distressed herself, and tried to help. Other empathy tests on rats had similar results: the free rats helped other rats escape all kinds of unpleasant and even painful situations. In one test, researcher Nobuya Sato investigated whether rats would rather eat chocolate (which they love) than help another rat—the rats chose to help their companions before eating the chocolate 50 to 80 percent of the time, demonstrating that the urge to help was at least as strong as their urge to eat a favorite treat.

Despite the bad reputation that humans have given rats—associating them with attributes like betrayal ("you dirty rat") and the meaningless pursuit of reward ("rat race")—it turns out that rats are unexpectedly compassionate and empathetic, and often value helping others more than receiving food rewards for themselves. Hopefully these discoveries will inspire people to treat rats more compassionately, especially in research.

Rats—who pass empathy tests with flying colors—seem to be one of the "good Samaritans" of the animal world—always ready to lend a helping paw.

Survival of the Kindest

Until recently, scientists believed that empathy was primarily a mammalian characteristic, but current research reveals that birds are capable of empathy, too. The evidence is particularly strong with ravens, an intelligent and socially complex species.

A raven typically socializes with a flock of friends for up to ten years before settling down with a mate, which led researchers Orlaith Fraser and Thomas Bugnyar to wonder if unpaired ravens might have empathetic friendships with one another during their "single" years. They studied thirteen captive juvenile ravens for twenty-three months, particularly watching for conflict behaviors. During this time, they observed 152 fights between the juvenile birds, noting the nature and intensity of the conflicts and who was the "aggressor" and "victim" in each. They also kept track of how "bystanders"—nearby flock members—behaved during and after witnessing the squabbles.

Specifically, Fraser and Bugnyar were watching to see if bystanders comforted the victims of fights with "affiliation" behaviors, such as sitting in close contact, preening, or beak-to-beak or beak-to-body touching. And sure enough, the bystander flock members consoled the victims. Sometimes the victims sought out affiliation with bystanders, but other times fellow flock members freely offered affiliation.

For a bystander raven to console a victim, he has to understand that the victim is distressed and then respond in ways to relieve that distress. This sensitivity to the emotional states of others is a simple form of empathy that was previously attributed only to mammals. It seems likely that as scientists continue to learn more about the emotional and cognitive capacities of birds, they will find empathy in many other avian species. They already have evidence for it in rooks, parrots, geese, and chickens.

It's also likely that empathy extends beyond fur and feathers. Based on crocodilian behavior, it looks like empathy might cross the reptilian divide. Female crocodiles respond to distress calls in hatchlings—both their own and those of other females. As for fish, studies have established that they respond to pain, experience stress, have long-term memory, and cooperate and reconcile with one another. So it is even possible that some species of fish—especially those with larger brains, such as sharks, skates, and rays, might have the capacity to empathize. Manta rays in particular are a strong candidate because of their cognitive abilities and social behavior.

Crocodiles might be cold-blooded, but they have warm-hearted traits: they play with toys and give one another piggy back rides, and the females respond empathetically not only to their own hatchlings but to those of other crocodiles, too. Crocodiles have even been known to form playful relationships with river otters and behave affectionately toward humans.

An increasing number of researchers believe that empathy may be a widespread capacity that developed millions of years ago, with its evolutionary roots in maternal behavior. Females of any species who are sensitive to the emotional and physical states of their young tend to respond to them more attentively, thereby increasing the chances that those offspring will survive and pass on their genes. For some species, evolution might favor empathy through "survival of the kindest."

REACHING ACROSS THE SPECIES DIVIDE

Empathy might not be hard to imagine among individuals belonging to the same species or among species that live together, such as dogs and people. In these examples, empathy benefits both parties. When rats are sensitive to and help one another, it is in the interest of the species as a whole; and when dogs comfort their human companions, they are likely to receive affection and food in return. But what about when empathetic behavior occurs between two completely unrelated species? Why would an animal reach across the species divide to help another?

Consider the crow and the cat. The story starts when Ann and Wally Collito, from Massachusetts, noticed a young kitten walking around the edge of

Bonobos appear to be able to understand the suffering of others and often take the initiative to comfort or reassure those in distress.

their property. A few days later, they saw a crow hanging around the kitten and worried that the crow might be hunting her. Instead they soon realized that the crow was "babysitting" the kitten, which they named Cassie. Over time, the couple witnessed (and videotaped) many instances of the crow, which they named Moses, feeding insects and worms to Cassie, grooming her, and keeping her safely off the nearby busy street. Though we can never know what Moses was thinking or feeling, the way he cared for Cassie certainly speaks of empathy and altruism.

In his book *Our Inner Ape*, Frans de Waal tells of Kuni, a bonobo housed at England's Twycross Zoo. When Kuni saw a starling fly into a pane of glass and fall to the ground, she immediately reacted to help the bird get airborne again. First, she tossed the bird into the air. When that didn't work, she climbed a tree with the bird, carefully unfolded her wings, and launched her into the air. The bird still hadn't recovered enough to fly and fell once again to the ground. Kuni climbed down and watched over the bird until she appeared ready to try to fly on her own.

Can there be any doubt that what Kuni displayed was a sophisticated sense of empathy and altruism? There was no reward or benefit for Kuni in trying to help the starling, so it's difficult to attribute her helpfulness to anything except the satisfaction of helping. Kuni also demonstrated a complex level of perspective taking: she had to think about what it's like to be a creature entirely different from herself (a bird) and imagine the kind of help this creature might need (resuming flight).

Unsurprisingly, the bonobo brain, like the human brain, contains spindle cells, which are associated with empathy and social interactions. As mentioned in chapter four, the brains of several species of whales also contain spindle cells. In 2009, one

Humpback whales risk injury to protect other species from orcas. Though we don't know why the humpbacks do this, more scientists are considering the possibility that these whales have altruistic motivations.

researcher witnessed puzzling humpback behavior that supports the idea that spindle cells might translate into whale empathy.

Marine ecologist Robert Pitman saw a group of orcas hunting a Wendell seal. The seal climbed onto a raft of drifting ice, presumably in an attempt to escape, but the orcas created a wave that knocked him back into the water. It looked like it was over for the seal—until a pair of humpbacks arrived. The seal swam toward the humpbacks, and as he got close, a fortuitous ocean wave lifted him right up onto the chest of one of the whales, who was floating on her back. As the seal started to slip off the whale's chest, the whale gently nudged him back on with her flipper. A short time later, when the marauding orcas had left, the seal made his way to the safety of another island of ice.

Curious, Pitman reached out to others who regularly observe humpbacks to find out if they had ever seen humpbacks come to another animal's defense. He received 115 descriptions of similar encounters, many accompanied by photographs and videos. In 90 percent of those encounters in which the animal under attack could be identified, it was not another humpback, so it was clear the whales were not defending one of their own. No one knows why humpbacks appear to sometimes come to the rescue of other animals. Although some scientists speculate that in the case Pitman witnessed, it may have been more about the humpbacks chasing off the orcas than defending the seal, that explanation still doesn't seem to account for the humpback's actions. Scientists have no real answers, but when we consider that the brains of these gentle giants contain spindle cells—the seat of emotions—it's hard not to give humpbacks the benefit of the doubt.

A final story of interspecies empathy involves Chantek, an orangutan who learned sign language. Once, he and his caregiver, Lyn Miles, were standing outside when it started to rain. Chantek picked up a scrap of fabric, tore it in half, and handed one half to Miles. He then held the other half above his head as if it were an umbrella, looked at Miles, and made the sign for "rain."

7. A SENSE OF THE SACRED

DEATH AND SPIRITUALITY

DEARLY DEPARTED

In her book *How Animals Grieve*, anthropologist Barbara King tells the story of Storm, a thoroughbred who suffered a severe injury and had to be put down. He was buried in the same field where he had spent his life grazing with his friends. When Storm's owner, Mary Stapleton, went to place flowers on the mounded earth that covered his remains, members of Storm's herd—his best buddies—stopped grazing and joined her, gathering in a circle around the mound. Other nearby horses, who were new to the farm and hadn't known Storm for very long, didn't join the circle. Those who did, however, stood in silence with Mary, their heads lowered.

Were Storm's friends grieving for him or engaging in some kind of ritual honoring their fallen friend? Although we will never know, there is increasing evidence that humans are not alone in understanding death, undertaking rituals honoring the dead, and experiencing sorrow at the loss of life. The way that many animals respond to death—from holding "funerals" and "wakes" to exhibiting depression and despair—suggests that they feel death's finality as deeply as we do.

> During a magpie "funeral," magpies placed blades of grass, pine needles, and twigs on the body of a fallen friend and stood still for a few seconds before flying away one by one.

We can surmise that animals understand death from the way they respond to the bodies of their mates, companions, and social-group members. Ethologist Marc Bekoff once saw a red fox bury her mate after a cougar had killed him. The fox covered him in soil and twigs, patted down the mound with her forepaws, and stood for a moment. She then walked away, with her tail down and ears laid back against her head, which is indicative of unhappiness in canids. Bekoff has also witnessed what could be called a magpie "funeral." One day he observed four magpies gathered around the body

Foxes, badgers, chimpanzees, elephants, and other animals have been observed burying the bodies of their family members or others in their social group.

of a dead magpie on the side of the road. They took turns gently touching its body. Then one magpie flew off, quickly returning with some grass, which he placed next to the body. Another magpie did the same, dropping pine needles and twigs on the body. Afterwards, all four magpies stood still for a few seconds before flying away one by one.

Science writer Jennifer Ackerman recounts a similar story of a dead crow encircled by twelve other crows, who were hopping around the body. After a few minutes, one of the crows left, returned with a small twig, dropped it on the body, and then flew away. One at a time, each of the other eleven crows did the same. Finally, only the body remained, covered in grass and twigs.

Gorillas have also been known to bury their dead by covering the body with twigs and leaves. Koko, the gorilla who knows sign language, once had an exchange about death with her trainer Penny Patterson. When Penny asked, "When do gorillas die?" Koko signed "trouble, old." When asked how gorillas feel when they die—happy, sad, or afraid—Koko signed "sleep," and when asked where gorillas go when they die, Koko signed "comfortable hole, bye." Koko's reference to holes in the context of death puzzled Patterson because no one had ever talked to Koko about burial, nor had she ever witnessed one. Patterson believes that burial may be an instinctive behavior for gorillas, based on her observations of gorillas raised in captivity (and therefore not exposed to wild gorilla behavior) burying dead birds they found in their enclosures.

Some species also appear to have "wakes." Gorillas and chimpanzees, for example, will quietly gather around a body and take turns sniffing and gently touching it. Captive gorillas have been observed holding the

Chimpanzees have very human-like responses to death. They tend to an ailing friend or family member by staying close and gently stroking them. After a chimpanzee dies, the friends and family of the deceased are often very subdued and behave in mournful ways.

hands of the deceased and laying their heads on the body. A video posted online in 2016 captures a similar reaction from a group of chimpanzees at the Chimfunshi Wildlife Orphanage Trust in Zambia. The resident chimps discovered the dead body of a member of their troop—called Thomas by the human caregivers—who had died of a respiratory illness near the fence surrounding the sanctuary. Researchers happened to be on the other side of the fence with a camera, so they videotaped what followed.

During the course of twenty minutes, twenty-two of the forty-three chimps in the compound gathered around, most sitting in silence, a few tenderly touching and sniffing Thomas's body, and one chimp even grooming the body. A chimp named Pan, who was an especially close friend with Thomas, smacked a disrespectful young chimp when he tried to move the body. During the "wake," all the chimps were quiet and even solemn, behaviors decidedly unusual for normally rambunctious chimps, especially since it was feeding time. But even food was not enough to lure Thomas's companions away from their friend. All of this behavior, because it was so ritualized, seems to be more about grieving than simple curiosity about a dead body.

Because of evidence that some animals seem to need time to grieve death, a growing number of zoo and farm staff have adopted the practice of leaving a body out for other animals to see and touch. In *How Animals Grieve*, King tells the story of two goats—Myrtle and Blondie—who were best friends. When Blondie died, she was at the vet's office, so Myrtle had no idea what had happened to her friend. When Blondie didn't return, Myrtle ran around the farm in a panic, looking for Blondie in all their favorite places and bleating loudly. After hours of witnessing Myrtle's anguish, the farmer retrieved Blondie's body and placed it where Myrtle could see it. Myrtle immediately went over to Blondie and sniffed her body, which finally settled her down. Over the course of a few days, she mostly stayed with the body, sometimes wandering off for a while, but always returning. Over time, her grief seemingly assuaged, Myrtle returned to her friend's body less and less, until the farmer was ultimately able to remove it.

Anyone who has spent time with goats will tell you that they are remarkable creatures—intelligent, curious, sociable, and fearless. Like dogs, they know their names, wag their tails, and develop close bonds with one another and their human caregivers. So it's no surprise to learn that they grieve when a close companion dies.

Evidence points to animals not only reacting immediately to death, but also remembering their long-dead. For example, wild elephants often stand vigil over a recently deceased herd member, stroking it for hours with their trunks, but they also appear to honor those who have been dead for a long time. They visit the sun-bleached skeletons and caress the larger bones and skulls. Even when encountering the carcass of an unrelated elephant, they often stroke the carcass before burying it under leaves and branches. Elephants also have been known to bury the bodies of other species, including humans.

When elephants encounter the bones or carcass of another elephant, they sniff and touch the remains. If they are in a group, they often stand quietly by the body and then gently touch and rub up against one another—gestures of consolation and reassurance.

In addition to ritual behavior, some animals exhibit strong emotional responses to death, just as humans do. When a companion or mate is lost, they will stop eating, stare into space, and withdraw from their group, or wail, howl, and act out. During their "wakes," chimps will sometimes break their silence to softly whimper or burst out in distress calls. After losing a companion, captive gorillas have been known to shriek and bang on walls or desperately try to revive the deceased.

Despite all these and so many other observations of mourning in the animal kingdom, skeptics assert that what appears to be grief may just be stress caused by a change in environment. But when a human loses a loved one, who would describe that person's grief in these terms? We would say the bereaved was sad because she had lost someone she cared about. If animals can develop bonds with one another—and we know they do—then why wouldn't they experience grief in response to the death of their mate or companion?

As humans, we have an inescapable awareness of death. From a relatively young age, we know that everyone dies eventually, including ourselves. Most scientists do not believe that animals have this kind of understanding of their mortality, but their more immediate sense of death suggests that they recognize its finality and suffer its pain. Grief unites us in a poignant and profound way.

SPIRITUALITY

Does animal awareness extend from death to some kind of overall sense of the sacred? Primatologist Jane Goodall thinks so. She has often told the story of the "waterfall dance," a ritual she has witnessed more than once among chimps gathered at the base of a giant, roaring waterfall. The chimps sway rhythmically, throw large rocks and branches, and swing over the water on hanging vines. They also behave

Based on their behavior, it's possible that chimpanzees and other apes may experience the power of nature as a kind of spiritual energy.

similarly during heavy rains in what is known as the "rain dance." Goodall speculates that when chimps perform these rituals, they might be experiencing something like awe in response to the power of nature.

Fire has elicited a similar response from chimps. Researchers Jill Pruetz and Thomas LaDuke watched as savanna chimpanzees in Senegal reacted uncharacteristically to fire, climbing trees to watch it rather than fleeing from it as other animals did. One male even performed a slow and exaggerated display—a "fire dance"—that seemed directed toward the fire.

There are other aspects of nature that appear to evoke a sense of the sacred in chimps. In 2016, a team of eighty scientists published a paper on chimpanzees behaving strangely toward hollowed-out trees at four field sites in West Africa. The chimps at each of these locations stacked stones inside the hollowed trunks of trees in a way that was superficially similar to ancient cairns discovered at human archeological sites. The chimps would later return to the tree, remove a stone from the pile, and hurl it at the tree. Laura Kehoe, one of the coauthors of the paper, speculated that this might be the first evidence of chimpanzees attributing special—perhaps even sacred—meaning to specific trees and behaving in ritualized ways. Other explanations were suggested, including the possibility that it was a kind of male dominance display. But that appears unlikely, as an adult female and juvenile chimp were observed engaging in the ritual, too. What the chimps are actually doing and thinking remains a mystery, but if humans were doing the same things, anthropologists might assume that the behavior reflected some sort of symbolic thinking and belief.

Chimps aren't the only primates to engage in behaviors that suggest they might possess a sense of the sacred. Primatologist Barbara Smuts lived with and studied baboons for two years, and one of her most memorable experiences involves what she came to call the "baboon sangha." A *sangha* is a Buddhist term meaning "assembly" and usually refers to a monastic community. As Smuts recounts the story, a group of baboons in Gombe were headed back to their sleeping trees when they came upon several still pools of water. Without any perceptible communication between the baboons, they all suddenly stopped traveling and sat on rocks at the edges of the pools. All were quiet—even the juveniles—as they sat and gazed at the water in what seemed to be a contemplative state. About half an hour later—again with no perceptible communication between

Charles Darwin believed that "the difference in mind between man and the higher animals, great as it is, certainly is one of degree and not of kind." So perhaps other primates, who have so much in common with humans, experience some kind of "sense of the sacred," too.

them—the baboons left the stream and resumed their journey in what Smuts described as "an almost sacramental procession." Though Smuts does not claim that this was an example of spirituality in baboons, she is willing to consider it.

> If animals can share with humans the capacity to feel joy, friendship, empathy, grief, and other emotions, perhaps they also share our capacity for numinous experiences evoked by the mystery and power of the natural world.

These and other ritualized behaviors have caused some scientists to consider the possibility that animals might have "spiritual feelings." Religious scholar Donovan Schaefer believes that human religion evolved not from a cognitive process, but rather from emotional and physical responses to nature. Schaefer believes that even though it is unlikely that animals experience a sense of divinity the same way that humans do, they may still be capable of sensing the "spiritual" energy in nature. Other scholars support this idea, believing that these ritualized primate behaviors may share similarities to the nascent religious behaviors of humans of the upper Paleolithic period (beginning about forty thousand years ago). So perhaps we are seeing the roots of human spirituality in our primate cousins.

With their ritualized behaviors, chimpanzees in particular are making it increasingly difficult to deny the possibility that animals might possess a sense of the sacred. Even the "sanghas" of the baboons is challenging to explain without allowing for such perceptions. Of course, these behaviors might be motivated by other factors, but given that there are so many other evolutionary continuities between animals and humans, why wouldn't there also be a spiritual continuity? If animals can share with humans the capacity to feel joy, friendship, empathy, grief, and other emotions, perhaps they also share our capacity for numinous experiences evoked by the mystery and power of the natural world.

PART TWO
MIND

Clearly, animals know more than we think, and think a great deal more than we know.

—*Irene M. Pepperberg*

8. TO KNOW THYSELF

awareness and identity

mirror, mirror

Once, after unsuccessfully begging for juice, Koko, the world-famous "talking" gorilla, settled for drinking water from a pan on the floor. She used a thick rubber tube as a straw to suck it up, and then, using sign language, described herself as a "sad elephant." In making an association between her makeshift straw and an elephant's trunk, Koko was using metaphor to comment self-referentially. Koko also signs to tell jokes; to blame others for something she did; to express emotions such as happiness, sadness, anger, fear, and love; to ask for hugs, treats, pets, and visitors; and to communicate her desire to have a baby. When looking in a mirror, Koko makes faces, checks out her teeth, and has even applied lipstick. When asked what she sees in the mirror, Koko signs "me," and when asked who she is, she answers, "Koko" and describes herself as a "fine animal gorilla."

Using lexigrams (visual symbols that represent words), the renowned bonobo, Kanzi, can ask to make a fire and roast marshmallows. (Remarkably, he is able to accomplish these two tasks on his own.) He also uses lexigrams to offer guests coffee, request outings, invite a human friend to play games, and even tell his caregivers to "be careful" when he sees them slip. And when asked if he is ready to do something that he has been waiting to do, he sometimes sarcastically replies "past ready." Like Koko, Kanzi also recognizes himself in a mirror, as well as in video. In fact, Kanzi used a video camera and monitor to practice blowing up balloons and bubble-gum bubbles. These were two activities he enjoyed but needed to practice, and the video setup offered him an opportunity to watch himself as he improved his technique.

Once, Chantek the orangutan looked up at the night sky, pointed at the moon, and asked his caregiver, "What is that?"

Another ape, the orangutan Chantek, uses sign language to request favorite foods, rides to restaurants, cage cleaning, and more. He understands the concept of money, and at one time he had an allowance—tokens—that he received in exchange for cleaning his enclosure and doing other chores. He used his allowance to "buy" himself treats, such as ice cream. Once, when looking up at the night sky, he pointed to the moon and asked his caregiver, "What is that?" Joining the ranks of Koko and Kanzi, Chantek recognized his reflection as his own and even described himself as an "orangutan person." When he was moved to a zoo with non-signing orangutans, he referred to them as "orange dogs." Chantek knew they were orangutans, so his trainers believe his "demeaning" description was intended to set himself apart from them.

"Talking apes" like Koko, Kanzi, Chantek, and a few others have made it possible for scientists to get a glimpse into a nonhuman animal's mind and to gather evidence of self-awareness. The ability to recognize oneself as separate from others—an aspect of self-awareness—is considered an indicator of intelligence. But apes that communicate a sense of self via sign language or lexigrams are rare exceptions in the animal kingdom, so scientists rely on other criteria to judge whether an animal is self-aware. One criterion is whether an animal passes the mirror test, a self-awareness assessment tool developed by the psychologist Gordon Gallop, Jr. in 1970.

Orangutans are not only self-aware—they also use tools, teach one another, plan ahead, engage in deception, and are capable of learning human sign language.

The test involves exposing an animal to a mirror and observing his reaction to determine whether he recognizes the reflection as his own. If an animal does not react to his reflection as his own, and instead ignores it, attacks it, or tries to engage it in play, he fails the test. He partially passes the mirror test if he behaves in a way that suggests he recognizes the reflection as an image of himself. The researcher then places a mark—using scentless paint, dye, or stickers—on an area of the animal's body that the animal cannot see without the mirror. To fully pass the test, the animal must recognize his reflection, notice the foreign mark in the reflection, and then try to examine the mark on his own body.

To date, eleven species (not including humans) have fully or partially passed the mirror test: ants, Asian elephants, bonobos, Bottlenose dolphins, chimpanzees, European magpies, gorillas, manta rays, orangutans, orcas, and rhesus macaques. Many of these animals behave in humanlike ways when in front of a mirror or video monitor. They may check out parts of their bodies by wiggling and shaking them; repeatedly open and close their mouths, stick their tongues out, make funny faces, and try to look at their backsides. One chimpanzee, Austin, who lived at the same research facility as Kanzi, fetched a flashlight to get a better look down his throat, an activity he initiated without any prior demonstration or prompting by his caregivers.

Magpies—known for their mischievous theft of shiny objects—are intelligent, social birds capable of reasoning, strategy, and foresight. They also play a game of hide-and-seek at a level comparable to that of preschool children: they take turns concealing themselves, peek out from their hiding places, and call out to their companion when they are ready to be found.

European magpies were the first non-mammalian species to pass the mirror test. Prior to this experiment, scientists believed that self-awareness arose from the neocortex, a brain structure found only in mammalian brains. Since birds don't have a neocortex, no one expected them to be self-aware. But when these corvids reacted to their reflections in the mirror, scientists had to change the way they thought about birds and brains. The magpies moved their heads, flapped their wings, and repositioned their bodies. And when they noticed the strange mark on their feathers in their reflection, they tried to peck and scratch at that place on their bodies. They passed with flying colors.

Another non-mammalian species that overturned conventional scientific thinking was a fish—the manta ray. In 2016, Csilla Ari, a researcher at the University of South Florida, filmed two giant manta rays swimming in a tank. The rays were filmed once without a mirror to record their normal behavior and once again with a mirror to see if they changed that behavior. When the mirror was present, the mantas displayed atypical behavior that indicated some measure of self-awareness: they circled in

Manta rays—who have one of the highest brain-to-body mass ratios of all fish—engage in inquisitive, play-like behavior and appear to be especially curious about humans.

front of the mirror, wiggled their fins, and blew bubbles, which is unusual behavior for a manta. They did not try to socially interact with their reflections, which would have suggested that they thought their reflections were other manta rays. The mantas were not marked, so they didn't have an opportunity to inspect themselves and fully pass the mirror test, but researchers believe that the mantas' responses strongly suggest that they are self-aware.

Even an insect—a species of forager ant—recently passed the test. Upon encountering the mirror, the ants turned their heads from side to side, flickered their antennae, and touched their reflections with their mouths. After the researchers painted tiny blue dots on their clypeus—a flat area above an ant's mouth—the ants noticed this change in their reflections and repeatedly rubbed their clypeus with a foreleg and touched it with an antenna in an apparent attempt to clean it.

That ants—tiny invertebrates—appear to have some measure of self-awareness is an invitation to scientists to consider the possibility that all animals might be self-aware, at least on some level. But better tests need to be devised. Although the mirror test is still considered the gold standard assessment, researchers are increasingly recognizing its limitations. For example, so far the only gorilla to pass the test is Koko. Other gorillas have failed, probably due to wild gorillas' aversion to direct eye contact, which is often interpreted as an aggressive gesture. This may be why most gorillas are naturally disinclined to stare at themselves in a mirror. Koko, as a result of her training and human enculturation, might have grown more comfortable with eye contact and was therefore able to engage the mirror long enough to recognize herself.

How widespread might self-awareness be? Recent evidence suggests that even forager ants have at least a rudimentary sense of self.

When looking for proof of self-awareness, researchers are realizing that they need to consider each animal species within the context of its unique way of sensing and responding to the world. Not taking behavioral and sensory differences into account could result in the false assumption that an animal lacks self-awareness. This even applies to human studies involving the mirror test. There was a widely held belief that most human children would pass the mirror self-awareness test by the time they were two. In reality, children from different cultures display different developmental patterns of self-awareness. Non-Western two-year-olds often fail the mirror test, and sometimes do not pass it until age six.

For example, in Kenya, of the eighty-two children (aged between eighteen months to seventy-two months) who were tested, only two passed. There was nothing wrong with the kids who failed—they were healthy, well-adjusted children. They simply were not used to mirrors and therefore behaved differently when placed in front of one, appearing uncomfortable when looking at their own reflections. Their failure to pass the test was not interpreted as evidence that they lacked self-awareness, but only that cultural differences must be taken into account. Similar considerations need to be applied when testing animals. Just as gorillas have taboos against eye contact, it is likely that other species might have other behavioral or perceptual characteristics that would impact how they respond to a mirror.

Dogs, for example, fail the mirror test, but most people wouldn't deny them a sense of self. Biologist Marc Bekoff has speculated that dogs fail the mirror test because the most important sense for dogs is smell rather than sight. They have roughly three hundred million olfactory receptors in their noses, compared to about six million in humans, and the part of their brains that processes smells is proportionately about forty times bigger than that of the human brain. Rather than asking dogs to recognize their own reflection, we should ask them to recognize their own scent—which is what Beckoff did.

In an experiment Bekoff nicknamed the "yellow snow test," he carefully collected samples of snow soaked with other dogs' urine and placed them along walking trails where his dog Jethro would encounter them. Jethro showed much less interest (exhibited by sniffing or urinating over the sample) in his own urine than in the urine of other dogs, suggesting that he had some sort of ability to recognize his own scent. Evolutionary biologist Roberto Cazzolla Gatti recently built upon Bekoff's research and developed the "Sniff Test of Self-Recognition" (STSR), which provided more evidence of self-awareness in dogs and further demonstrates that self-awareness may not be as rare as once thought. Scientists just need to figure out how to look for it.

It wasn't until scientists considered the way canids perceive the world—primarily through scent—that they were able to develop a dog-appropriate self-recognition test.

There's no doubt that the mirror test has been a helpful step in establishing the idea of self-awareness in animals. However, like the magic mirrors of fairy tales, it has also revealed at least one unexpected truth: humans may well be the most self-aware of all, but there are still plenty of times we fail to see our own biases and the ways they shape our perceptions.

IT TAKES ONE TO KNOW ONE

Using sign language, Koko the gorilla has been known to tell a lie—such as denying she did something against the rules—to avoid consequences. When she lies, she is demonstrating a complex understanding of communication: she understands others have expectations (such as "follow the rules") and emotions (such as anger when she doesn't follow the rules). It's also likely that she believes the other person might not have knowledge of the truth or could be persuaded to doubt their knowledge.

Koko's lies demonstrate what scientists call a "theory of mind," which is another criterion for self-awareness. It is the ability to attribute mental states to oneself and others. It allows a human or nonhuman animal to recognize its mind as separate from other minds and to understand that others have their own mental states, such as intentions, beliefs, knowledge, desires, and perspectives. Theory of mind is also called "perspective taking" because it involves imagining the perspective of another. It can foster empathy, but when negative intent is assumed it can also elicit distrust.

So far, studies have shown that gorillas, chimpanzees, and other primates demonstrate a theory of mind, as do other mammals, such as dogs and pigs. Scientists have also observed theory of mind behavior in crows, ravens, jays, and other corvid species. Many corvids cache their food for future use and retrieve it as needed. But if they notice that other birds are watching them while they hide their food, they will return to their stash later and move it to a different location. Their behavior suggests they understand that other birds have minds—and intent. As it turns out, the corvids most likely to re-cache their food out of fear of theft are also the ones most likely to steal another's cache. Clearly, it takes one to know one.

Answering by Name

Recognizing themselves in the mirror and understanding that their mind is separate and distinct from the minds of others are two ways animals demonstrate self-awareness. Another is having a name by which they identify and distinguish themselves.

When a bottlenose dolphin is just a few months old, it develops a unique vocalization—a pattern of notes that becomes its "signature whistle." No two signature whistles are alike. It appears that juveniles develop their signature whistles by listening to, learning from, and modifying the calls of other dolphins, usually family members. Dolphins mostly vocalize their own whistle, but sometimes mimic one another's whistles.

Marine biologists Stephanie L. King and Vincent M. Janik wondered if these signature whistles might be functioning as "names" that dolphins use to announce themselves and address one another. To find out, they carried out two studies. In the first, researchers recorded the whistles of captive and wild dolphins in Florida, and discovered that only dolphins with close social bonds recognized and mimicked each other's signature whistles and that the mimicking tended to occur when the bonded dolphins were separated. It seemed like the dolphins were calling out their friends' signature whistles in the same way that humans call out the names of family members and friends when they are looking for them or trying to get their attention.

In their second study, researchers followed different groups of wild dolphins off the coast of Scotland,

Dolphins call and respond to one another by producing unique sounds—known as signature whistles—that function as names. Having names isn't just useful in social situations. Dolphins might use their signature whistles to call out to friends for help or to invite them to a good fishing spot.

recording individual dolphin's signature whistles. Using software, they then created "synthetic" versions of the whistles, which were altered in order to remove the unique vocal qualities that identified the whistle as coming from a specific dolphin. For example, imagine listening to a recording of your mother's voice as she called your name. You would almost instantly recognize both your name and your mother's voice. Now, imagine manipulating that recording by substituting a robot voice. You would still recognize your name, but you wouldn't recognize the voice as coming from your mother.

By neutralizing the vocal tones, researchers could be certain that when the dolphins heard the recordings, they would hear the message—or "name"—without being able to identify the speaker. If they recognized the "name" but not the "speaker," would they still react? Doing so would mean it was the "name" that actually carried meaning for the dolphins, not the familiar voice.

King and Janik observed the dolphins' responses to three kinds of synthetic signature whistles: their own signature whistle, the whistles of dolphin "friends," and the whistles of dolphin "strangers." When a dolphin heard its own signature whistle, its "name," it whistled back. In fact, some dolphins, after hearing their own signature whistle, approached the source of the sound—the research boat—as if to say, "Here I am! You called?" In contrast, hearing the synthetic signature whistles of friends and strangers didn't elicit much of a response, just as hearing a digital voice say "Jane Doe" wouldn't grab your attention, unless you happened to be named Jane Doe.

These researchers speculate that signature whistles are used in similar ways to how humans use names: to initiate contact ("My name is Jack. What's your name?"), to help dolphins find each other ("Hey Jack, where are you?"), and to help calves recognize their mothers ("Mom, is that you?"). More research is needed to determine whether two or more dolphins might use signature whistles to refer to dolphins who are not present ("Did you hear what happened to Jack?").

In addition to passing the mirror tests, the dolphins' capacity to make and recognize individual "names" provides more evidence that they have both a sense of self and the ability to distinguish self from other.

Although researchers are still far from being able to converse with dolphins, at least now they may be able to open communication by greeting them by name.

Straight from the Parrot's Mouth

Other animals, such as wild parrots, have "names," too. Like a dolphin's signature whistle, each parrot appears to have what scientists call a "signature contact call," a unique chirp used to identify himself to others. Cornell University ornithologist Karl Berg wanted to know where these signature contact calls, or names, came from: Were they genetically encoded or were they derived from the nest and learned through mimicry? In other words, he wanted to know if the chicks were born with their names or if the parents somehow assigned names to their chicks. (Mom: How about "Tango"? Dad: What's wrong with "Tweety"?)

Berg decided to study the green-rumped parrotlet to seek an answer to this nature-versus-nurture question. He set up video cameras to monitor the activity of seventeen parrotlet nests, and he recorded the calls of all the parrot parents. He then switched eggs in nine of the nests, so that those chicks would be raised by "foster parents." If these fostered chicks grew up with a signature contact call that sounded like their biological parents instead of their foster parents, then he would know the calls were encoded in their genes. But if the calls sounded more like their foster parents, this would mean the chicks did not inherit their signature contact calls, but learned them instead.

Humans aren't the only animals who name their children. Green-rumped parrotlets do so as well. When their chicks hatch, the parents chirp a unique call for each hatchling that becomes the chick's "name." Credit: Creative Commons BY-SA 4.0, Jam.mohd

After a two-month period, Berg used spectrographic software to analyze the calls and, sure enough, mom and dad were indeed naming their chicks. The parents chirped a different call at each chick, which the chick learned to chirp back, using their mimicry skills. After a few weeks of practice, each chick had learned his or her own signature contact call, which definitely bore a family resemblance to the signature contact calls of the parents who raised them, regardless of whether those parents were their biological or foster parents. The chicks also learned the signature contact calls of their parents and siblings, and the whole family used these "names" to communicate with one another.

"It's Christmas. That's what's happening. That's what it's all about. I love Pucky. I love everyone."
—Puck the parakeet, one Christmas morning

Some scientists believe that these calls—and the way parrots use them—suggest that parrots have a sense of self. Of course, most people who have lived with parrots would tell you that there's no question they have self-awareness. Take Puck, for example, a parakeet who made it into the 1995 *Guinness Book of World Records* as "the bird with the largest vocabulary in the world." Puck knew 1,728 words, and often created his own phrases, which he used appropriately. Like many parrots and parakeets, he sometimes talked to himself. One Christmas morning, Puck had a moment to himself in the living room when his caregivers heard him say, "It's Christmas. That's what's happening. That's what it's all about. I love Pucky. I love everyone." Perhaps that charming soliloquy was just well-timed mimicry on Puck's part. But based on what scientists are learning about parrot intelligence, it might be best to place your bets on self-awareness as a more likely explanation.

There are 356 known species of parrot, including macaws, cockatoos, and parakeets. Parrots are known for their intelligence, mimicry skills, playfulness, and mischief. Though parrots have not passed the mirror test, it is likely they have a sense of self based on other behaviors.

9. IF WE COULD TALK TO THE ANIMALS

LANGUAGE

TALK OF THE TOWN

In an episode of the popular nineties television series *Friends*, Monica asks Phoebe, "Do you think that your favorite animal says much about you?" Phoebe replies, "What? You mean behind my back?" If the prairie dog was Phoebe's favorite animal, then the joke's on us because prairie dogs appear to be doing just that—talking about us behind our backs.

Prairie dogs are rabbit-sized social rodents native to the grasslands of central and western North America. They live in complex underground tunnel and chamber systems with separate rooms for tending to their young, sleeping, food storage, and elimination. Next to their burrow exit is an area they use as a "listening post" to listen for alarm calls from their neighbors and raise the alarm themselves if they spot a predator. Their tightly knit family groups share food, groom one another, and greet by kissing on the mouth and nuzzling. Families are often grouped into "wards," which are like neighborhoods; and wards, in turn, are grouped into colonies, which are aptly called "towns."

If prairie dogs are starting to sound like people with their chambered dwellings, family groups, neighborhoods, and towns, wait until you hear about their "conversational skills." Animal behaviorist Con Slobodchikoff has been studying one species of prairie dogs—Gunnison's prairie dog—for more than thirty years. When he first began to study them, he noticed that when prairie dogs spotted a predator, they shouted warning calls, which were then picked up and repeated by others throughout the colony in a kind of emergency broadcast system. But the prairie dogs hearing the warning didn't always respond the same way, calling into question whether the call was a generic warning of "Danger!" or something more specific. Sometimes the alerted prairie dogs immediately retreated into their burrows; at other times they ran to the edge of their burrows and, instead of diving in, stood upright and watched. At still other times, they just stopped where they were out in the open and stood motionless, as if weighing their options. How did the prairie dogs know what kind of response to make when they couldn't see the predator, and therefore couldn't gauge their risk level?

To figure out what was going on, Slobodchikoff set up a recording system. Whenever a predator—typically a coyote, dog, hawk, or human—passed through the colony, he recorded their warning calls. He then used a software program to analyze the frequencies of the calls and discovered that the prairie dogs weren't just shouting out a generic warning cry—they were using different calls to specify different types of predators. What's more, as Slobodchikoff studied them more, he noticed subtle variations in the call

Prairie dogs have developed their own version of an "emergency broadcast system" that delivers warnings about specific predators.

One of the predators that prairie dogs "talk about" is humans. These chatty rodents have an alarm call by which they not only announce an approaching human, but also what he looks like. They even appear to have a call for "gun," which they use in combination with their call for "human."

for "human." He began to suspect that the prairie dogs weren't just calling out "human," but might be describing the human, too.

To test his hypothesis, Slobodchikoff recruited four human volunteers and dressed them in the same way except for the colors of their t-shirts, which were blue, green, yellow, and gray. One at a time, the volunteers walked through the colony while Slobodchikoff recorded the prairie dog calls. He then had the volunteers change into a different colored shirt and walk through again. He repeated this test until each volunteer had worn shirts of all four colors. After analyzing each call, he discovered that the calls consistently changed when the color of the volunteer's shirt changed. Slobodchikoff's hunch was right: the prairie dogs appeared to be describing the person walking through their colony.

> Prairie dogs can communicate surprisingly specific information, such as "small human wearing blue shirt walking slowly."

Slobodchikoff's research revealed even more startling details of these warning calls, all of which suggested that prairie dog communication has a complex structure, with different sounds functioning as different parts of speech. They have, loosely speaking, a kind of grammar. In his book *Chasing Dr. Dolittle: Learning the Language of Animals*, Slobodchikoff explains, "They have parts of their calls that are noun-like: human, coyote, dog, hawk. They also have parts that are adjective-like: yellow, blue, green, big, small. And they have verb-like and adverb-like parts: running fast, walking slowly." The prairie dogs were communicating surprisingly specific information, such as "small human wearing blue shirt walking slowly." They even appear to have a call for "gun," which they would use in combination with the call for "human."

Slobodchikoff has decoded more than one hundred prairie dog "words" that are used in the context of warnings. Of course, prairie dogs don't only vocalize to warn one another; they also vocalize socially in family groups and neighborhoods. However, we don't know what prairie dogs are talking about here because social chatter lacks context—it doesn't lead to any perceivable change in behavior, such as diving into a burrow. But based on what Slobodchikoff and others have already learned, it is likely that even their social chatter has meaning.

"Prairie dog" isn't the only animal language that scientists have started to decode. Technology is enabling researchers to study and translate the calls of bats, chickens, dolphins, elephants, gibbons, pandas, parrots, and other species. Slobodchikoff believes that soon we will have hand-held translation devices that allow back and forth communication between humans and animals. Once we start having "conversations" with animals, will we change the way we treat them? If we hear animals express fear and contentment, ask one another to play, or talk about their young, will it inspire us to put more conservation and preservation measures into practice? Will we give them something good to say about us?

YIPS, CHIRPS, AND BARKS: DECODING PRAIRIE DOG LANGUAGE

Gunnison's prairie dog language is the most sophisticated animal language decoded to date. Dr. Con Slobodchikoff has decoded over one hundred of Gunnison's prairie dog's yips, chirps, barks, and other vocalizations. Here are a few of the prairie dog's labels, or "words," organized by how they appear to function within prairie dog "grammar."

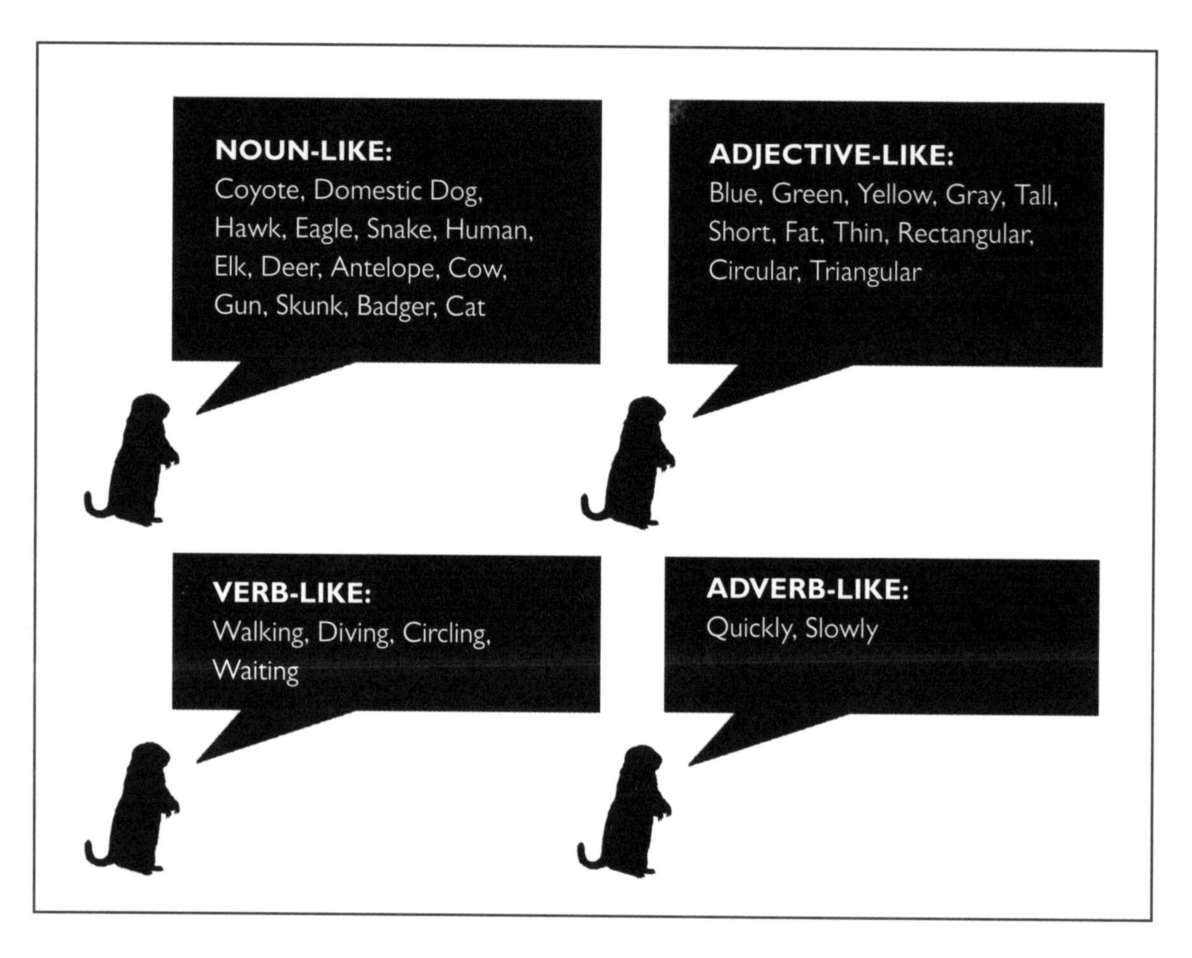

EAVESDROPPING OWLS

Imagine that you are a western burrowing owl, a small, ground-dwelling raptor that hangs out in the same habitats as prairie dogs. You need a place to nest, so you decide to move into an empty prairie dog burrow. Prairie dog burrows have a lot to offer: they are ready-made, so you don't need to dig your own; when it's time to nest, you have the option of expanding into nearby burrows as your family grows; and adjacent burrows serve as alternate escape routes should a predator arrive at the front door.

Those are all excellent reasons to move into a burrow in a nice, quiet, abandoned prairie dog town. But most burrowing owls don't move into burrows located in abandoned towns; they move into burrows in active, noisy prairie dog towns, where the chatty rodents are constantly shouting messages to one another.

If you were a burrowing owl with nesting in your future, why would you choose to make your home in a bustling metropolis of oversized rodents? One reason might be to reduce your chances of becoming someone's dinner. Coyotes and badgers prey on both burrowing owls and prairie dogs, so you might seek safety in numbers by hiding out in a crowd.

However, researchers Rebecca D. Bryan and Michael B. Wunder had a different hypothesis about why the owls prefer living with prairie dogs: to eavesdrop. Just as people learn from observing other people, animals learn from observing other animals, including different species. Prairie dogs—by broadcasting alarm calls—offer potentially valuable information about who is lurking in the neighborhood, and Bryan and Wunder believe the owls were listening to them.

Some animals, such as burrowing owls, learn to listen in on another species' communications and often gain valuable information about what's happening in the neighborhood.

To test for eavesdropping, they recorded the black-tailed prairie dog call for "snake," since rattlesnakes and bull snakes prey on burrowing owl chicks and eggs. Researchers then played back the prairie-dog "snake" call, as well as recordings of two control sounds—cattle mooing and an airplane engine—and compared the owls' responses to the three sounds. As predicted, compared to the mooing and the airplane, the owls reacted to the prairie dogs' snake calls with much more vigilant behaviors: they stood in alert poses; turned their heads to look around; walked, ran, or flew; bobbed their heads; and vocalized their own alarm calls. The results suggested that owls were eavesdropping on the prairie dogs and understood at least some of their language.

WHALES SPEAKING DOLPHIN AND DOLPHINS DREAMING IN WHALE

It's one thing for an animal to understand warning calls and eavesdrop on another species for its own protection, but it's another thing entirely to learn to speak another species' language.

Researcher Whitney Musser and Hubbs-SeaWorld Research Institute scientist Ann Bowles found that captive orcas living with captive bottlenose dolphins changed their vocalizations—duration, pitch, and pattern—to match the sounds made by their dolphin cohabitants. Was this evidence that the whales were learning to "speak dolphin?"

The research team made and compared two kinds of recordings: the vocalizations of captive orcas housed with bottlenose dolphins and those made by orcas who were housed separately. The vocalizations of the former had morphed into more dolphin-like sounds (compared to the latter). One whale even learned a new vocalization—a chirping sequence—that dolphins had learned from their human trainers before they were housed with the whales. Clearly the whales were learning new vocalizations from their friends.

Groups of orcas have their own local culture that shapes their diet, play activities, mate selection, and even their vocalizations. Large groups—clans—have vocalizations that differ as much as human languages differ. Smaller groups—pods—have dialects that are more analogous to a southern drawl or Boston accent. Wild orcas are able to learn the local dialects of other clans, so perhaps it's not that surprising that in captivity they learned to "speak dolphin."

Meanwhile, halfway around the world, Péos, Mininos, Cécil, Teha, and Amtan—performing dolphins at the Planète Sauvage Dolphinarium in Port-Saint-Père, France—were recorded speaking in their sleep. However, these dolphins weren't speaking dolphin, they were speaking humpback whale! Because these dolphins were born in captivity and had never lived in the sea where they might have encountered humpback whales, researcher Dorothee Kremers and her colleagues speculated that the dolphins might have been dreaming about their performance routine. The soundtrack played during the Dolphinarium show included humpback whale sounds, so it is possible the dolphins were mimicking that part of the soundtrack while "rehearsing" the performance in their sleep. These scientists have a lot more research to do before they can draw any conclusions, but they are extremely intrigued by these first investigations into the sleep and dream states of dolphins.

CHASING WORDS

What if our canine pals were not only good at keeping us company and following simple commands, but also happened to have one of the most impressive animal minds?

Retired psychology professor John Pilley and his colleague Alliston Reid explored the potential of canine intelligence by training Pilley's border collie, Chaser, to associate words with toys. Pilley wanted to know just how large a dog's vocabulary for human words could be. He had read about a border collie named Rico who knew two hundred words for toys, and he hoped Chaser could beat Rico's record. It didn't take long for Chaser to learn two hundred words, so Pilley kept teaching her, and Chaser kept learning. After Chaser had learned the names of 1,022 toys (balls, Frisbees™, stuffed toys, and plastic toys), Pilley stopped teaching her new words due to time constraints and his interest in exploring other aspects of Chaser's abilities. Pilley still doesn't know just how large a dog's vocabulary can be, but he knows Chaser's memory is better than his. In fact, both Pilley and Reid readily admitted that they could not remember all the names of Chaser's toys themselves, so they wrote the names directly on the toys with permanent markers to help them remember.

Dogs are wired to pay attention to humans, which is one of the reasons that dogs like Chaser can learn so many words. In fact, recent research suggests that when it comes to cooperative communication skills, such as the ability to follow a pointing finger or human gaze, dogs outperform chimpanzees and are comparable to human toddlers.

After teaching Chaser a vocabulary three times greater than the average toddler's, Pilley decided to find out if Chaser could tell the difference between nouns (the names of toys) and verbs (commands). He taught Chaser three commands—nosing, pawing, and taking. He and Reid then randomly mixed up the names of the toys with the commands. Amazingly, Chaser understood right from the start. As one of many examples, she differentiated the commands "Nose Blue," "Paw Blue," and "Take Blue"—three different actions taken in relation to the same toy.

Ever inspired by Chaser's genius, Pilley now turned his attention to categories. Chaser already knew that her playthings belonged to a set of objects called "toys," but could she learn that within this large set of toys there were subsets of kinds of toys, such as "balls" and "Frisbees™"? It turns out Chaser could differentiate and categorize based on qualities. She understood that the word "ball" referred to round objects that roll and/or bounce, and "Frisbee™" meant any spinning disc or ring-like object.

Finally, Pilley set up an experiment to find out if Chaser could infer the name of an object through the process of elimination. He added a new toy to Chaser's collection and gave it a new name. When he asked her to get "X," a name she had never heard before, Chaser quickly figured out that "X" was the new toy and brought it to Pilley.

Is Chaser special? Can other dogs understand language to this extent? The probable answer is yes on both counts: Chaser is special (as is John Pilley), and other dogs might be as smart as Chaser. Pilley's work with Chaser, along with that of other researchers with other dogs, has inspired more scientists to study canine cognition. Brian Hare, the author of *The Genius of Dogs*, founded a website called Dognition (www.dognition.com) that offers a web-based app allowing people to explore, test, and record their dog's intelligence. Hare hopes to use the crowd-sourced data from the website to expand our understanding of just how smart dogs might be.

The Dream of a Common Language: Teaching Apes to Talk

Being able to talk to animals is one of humankind's most ancient dreams—it reaches back to myths, shamanic traditions, and fables, and it lingers on in children's storybooks and films. Within the last sixty years, the dream of talking to animals has entered the domain of science.

We share approximately 99 percent of our DNA with chimpanzees and bonobos, 98 percent with gorillas, and 97 percent with orangutans, so apes might seem to be the best candidates for language learning. But anatomical differences in their tongues and voice boxes make it difficult for them to vocalize human language sounds. In a language research project that began in the late 1940s, Keith and Catherine Hayes tried to teach a chimpanzee named Viki to vocalize English words, but Viki just barely managed to learn four words, and only with assistance. After Viki's failure to learn to speak, scientists focused primarily on teaching apes sign language, although there are a couple of recent studies that are revisiting the possibility of teaching apes to vocalize.

Early Attempts

In the 1960s, psychologists Allen and Beatrix Gardner were inspired by the way wild chimps naturally gesture to one another, so they taught Washoe, a chimpanzee, how to use American Sign Language. Washoe was the first chimpanzee to learn sign language and mastered about 250 signs that she used for requests, such as asking for food, toys, tickles, and hugs, and for labeling objects in her environment. She could put together simple phrases, such as "Gimmie Sweet" and "You Me Go Out Hurry." She could also use signs to comment on something or someone. For example, she used the sign "dirty" as both an adjective (to describe soiled objects) and an insult (when her trainer refused her requests). Washoe also taught approximately fifty signs to her adopted son, Loulis, without any prompting from researchers.

"Darn darn floor bad bite. Trouble trouble."
—Koko, the gorilla
(signing after she experienced an earthquake)

Koko

Excitement over Washoe inspired additional studies, and the 1970s became the golden age of ape-language research. Perhaps the most famous "talking ape" from that era is the gorilla Koko. Psychologist Penny Patterson began teaching one-year-old Koko sign language in 1972. Forty-four years later, Koko knows more than one thousand signs and understands two thousand words of spoken English. She has communicated her preferences for everything from food to television personalities, shared feelings of joy and sadness, lied to try to avoid consequences, and even made jokes. Her vocabulary is comparable to that of a three-year-old child, and she often strings words together in ways that reveal creativity and a surprising depth of understanding. For example, after experiencing an earthquake, Koko signed, "Darn darn floor bad bite. Trouble trouble."

Chantek

In the 1970s, anthropologist Lyn Miles began teaching an orangutan named Chantek how to sign. Chantek learned 150 signs over eight years, and also understood many words of spoken English. Chantek used short strings of signs to make phrases, and one of his most interesting phrases expressed his wish for privacy: when he wanted a confidential exchange, he would sign "secret" and make his gestures small in an effort to conceal them. If people were visiting and Chantek wanted a moment alone with Miles, he would sign "I you talk," while pointing to a location away from the visitors. Chantek also seemed to understand that words could have multiple meanings, as when he signed "bad" after biting into a radish, as well as when encountering a dead bird.

Nim Chimpsky

Around the same time that Patterson was working with Koko, behavioral psychologist Herbert Terrance initiated a sign language project with a chimp named Nim Chimpsky (after the famous linguist Noam Chomsky). Nim learned 125 signs for objects and actions, and like Washoe, Nim could put together simple phrases, such as "Tickle me Nim play" and "Banana me eat."

However, by the end of the study, Terrance believed that Nim's use of signs was often the result of unintentional cueing by trainers rather than actual language use. Terrance also believed that just because apes can be taught to label objects correctly doesn't mean they have learned how to use language. According to Terrance, object labeling by itself does not meet the criteria for true language use, which requires understanding that the same words can mean different things when arranged in different orders. For example, the sentences "chimp eats banana" and "banana eats chimp" use the same set of words, but their word order results in very different meanings. He did not see any proof that Nim—or other apes in other studies—understood this critical aspect of language. Even though many researchers disagreed with Terrance's critique, many more accepted it; consequently, interest in ape language research declined.

Kanzi

In the early 1980s psychologist Sue Savage-Rumbaugh started working with bonobos (also known as pygmy chimpanzees). Savage-Rumbaugh addressed many of Terrance's concerns by using lexigrams—visual symbols that represent words—rather than sign language. By avoiding sign language and the possibility of unintentional nonverbal clues, Savage-Rumbaugh and her team demonstrated that the bonobos in her study were truly learning the meaning of words and not just responding to cueing from trainers.

Scientist Sue Savage Rumbaugh reported that Kanzi, like other "talking apes," sometimes combined signs to invent his own terms. For example, his board did not have a lexigram for "kale," so he pointed to "slow" and "lettuce" to represent "kale" because it took him a long time to chew kale. Credit: Creative Commons BY-SA 4.0, William H. Calvin, PhD

Savage-Rumbaugh has worked especially closely with one bonobo in particular, Kanzi. As an infant, Kanzi often accompanied his adoptive bonobo mother, Matata, to her lexigram training sessions. One day Kanzi spontaneously began to use the lexigrams, demonstrating that apes can learn language naturalistically rather than only through direct training. After years of training, Kanzi now knows approximately five hundred lexigrams and three thousand words of spoken English. Kanzi has the distinction of being the animal who understands more human language than any other animal in the world.

When Kanzi hears a spoken word or question asked by an invisible interrogator through headphones, he chooses the correct lexigram(s) in response. He also chooses lexigrams to communicate his own requests and, on occasion, to make comments. The lexigram board—which is wired to a computer—vocalizes out loud each lexigram that Kanzi presses. Kanzi also has a non-electronic version of his lexigram board—a portable laminated poster that can be carried offsite. Sometimes he also uses tablets with specially designed apps in place of the old lexigram board.

In Savage-Rumbaugh's more recent work with Kanzi, she reports that he is developing an understanding of the very basic features of a simple grammar, or "protogrammar." For example, Savage-Rumbaugh asked Kanzi a question he had not heard before: "Can you make the dog bite the snake?" Kanzi responded by searching his toys until he found a toy dog and a toy snake. He then put the snake in the dog's mouth and, using his thumb and finger, closed the dog's mouth over the snake. By making the dog bite the snake instead of making the snake bite the dog, he showed a basic understanding that word order matters.

In the roughly sixty years that ape-language research has been going on, all four species of great apes have learned how to "talk" to humans at a basic level. They have learned how to ask for specific foods, toys, and activities, and they have commented on events in their lives. They have learned to understand our requests to do all sorts of things that are unfamiliar in the world of wild apes, such as sitting at a computer while wearing headphones and pressing keys. Despite all that has been accomplished, skeptics still assert that the interpretations of ape-language studies are overly optimistic and reveal more about the researchers than the apes. Critics claim these researchers so desperately want to believe in the "fairy tale" about talking animals that they selectively interpret their research results to support their fantasy. For their part, many of the ape-language researchers claim the skeptics are so entrenched in their bias for human superiority that they have little ability to assess the research fairly.

Most scientists fall somewhere in the middle. They believe that even though ape-language studies leave room for interpretation, they provide a valuable and intriguing window into the minds of apes and the nature of language. Through these studies, scientists have at the very least learned that apes have self-awareness, imagination, a sense of humor, and a wide range of nuanced emotions. Even if apes never acquire human-like language in all its grammatical complexity, the fact that they can now reach across the linguistic divide and request a private conversation, ask to make a campfire and roast marshmallows, or comment on an earthquake is amazing in itself.

Chimpanzees have their own natural "sign language" in the wild. Scientists have identified dozens of different gestures that chimpanzees use to communicate with one another. For example, showing the sole of the foot to another means "climb onto my back." A mother chimpanzee might present this gesture to a whimpering juvenile, who then knows it's okay to get a ride from mom.

WORDSMITHING APES

Some of the apes that have been taught human language—various forms of sign language or lexigrams—have coined their own terms by combining signs in original and creative ways. These terms reflect their cleverness, creativity, and sometimes even poetic sensibility. Here are a few of the terms invented by a wordsmithing chimpanzee, bonobo, gorilla, or orangutan.

COINED BY KANZI, A BONOBO

Slow Lettuce for kale (because it takes Kanzi longer to chew kale than lettuce)
Big Water for a flood
Potato Surprise for potato chips

COINED BY KOKO, A GORILLA

Finger Bracelet for a ring
Bottle Match for a lighter
Eye Hat for a facemask
White Tiger for a zebra
Elephant Baby for a Pinocchio doll

COINED BY MICHAEL, A GORILLA

Orange Flower Sauce for nectarine yogurt
Bottle Necklace for a six-pack can holder
Bean Balls for peas

COINED BY CHANTEK, AN ORANGUTAN

Tomato Toothpaste for ketchup
Cheese Meat Bread for Big Mac®
Eye Drink for contact lens solution
Key Man for zookeeper

COINED BY MOJA, A GORILLA

Metal Cup Drink for a thermos
Metal Hot for a lighter

COINED BY WASHOE, A CHIMPANZEE

Water Bird for a swan

COINED BY LUCY, A CHIMPANZEE

Candy Drink for watermelon
Cry Hurt Fruit for radish

10. COUNT THEM IN

NUMERICAL COGNITION

DOES A BEAR COUNT IN THE WOODS?

Like people, bears can stand upright on the soles of their feet and sit on their tailbones with their legs stretched out in front of them. Using their front paws in hand-like ways, they can gather fruits and nuts, fashion backscratchers from branches, and throw rocks, snowballs, and other objects. Bears hum when content, snore in their sleep, and smile or laugh when relaxing, meeting a friend, playing, or watching other bears play. In addition to all of these traits and behaviors, scientists have discovered another commonality between people and bears: they can "count." Though we do not yet know for certain that bears count in the woods—in the wild—it is reasonable to assume they do, because they can count in captivity.

Bears have large and complex brains and an intelligence that compares with that of higher primates. In captivity, they have learned an incredible repertoire of tricks, such as riding bicycles and playing musical instruments. In the wild, bears are known for their clever and cunning natures and have even been observed covering their tracks to avoid being followed by humans.

Psychologist Jennifer Vonk was doing primate research at a zoo in Mobile, Alabama, where she had taught a chimpanzee to use a touch-screen computer. During her research, she met three American black bear siblings—named Bella, Brutus, and Dusty—who were also housed there. She wondered if the bears, like the chimp, could be taught to use a touch screen. She knew wild bears were clever and known for their abilities to get into bear-proof trashcans, open door latches, and manipulate other barriers to reach food. Additionally, captive bears are easy to train and have been taught to ride bikes, roller skate, play musical instruments, and engage in other complex tasks. So how hard could it be to get them to use a touch screen?

Vonk taught the bears to touch the screen with their noses or tongues, which is much gentler on hardware than bear claws. Before long, the bears figured out that touching the screen at the right time—in response to specific images—earned them food rewards. After the bears learned how to use the touch screen (pretty amazing in itself), Vonk and her research partner, Michael J. Beran, set up an experiment to test whether the bears could "count."

First they trained the bears to discriminate between two sets of dots, each containing a different number of dots. Brutus had to learn to choose the larger set, whereas Dusty and Bella had to learn to choose the smaller set. When the bears touched the correct set, they heard a melody and received an edible treat. When they chose the wrong set, the computer buzzed and no treat was given. With practice, all three bears seemed to grasp the idea.

Once the bears got the hang of it, the real testing began. During each test session, the bears were once again presented with two sets of dots. Each set contained a randomly chosen number of dots—between one to ten—and one set was always numerically larger than the other. To make sure that the bears could tell the difference between sets of dots based on quantity rather than density, the researchers varied the area, size, and arrangement of the dots. They also included tests in which the dots were animated and moved around the screen.

All three bears performed well and were able to figure out—to varying degrees—how to select the set with either the larger or smaller number of dots. Brutus did especially well, demonstrating that he could distinguish between larger and smaller quantities even when the researchers distributed smaller quantities over larger areas. Though the researchers do not know for certain that the bears were "counting," they were definitely distinguishing between quantities.

So why would a bear—or any animal—have this ability? After all, they don't need to file taxes or take algebra tests. But many animals need to be able to tell if one quantity is smaller or larger than another. For example, if you are a species of fish that prefers safety in numbers, you need to be able to discern which shoal is larger, and therefore safer. And if you are a bear that can judge which foraging area has more berries, you're more likely to nap with a full belly.

Counting Crows

Since at least the eighteenth century, farmers and hunters have claimed that crows are capable of counting. Now, scientists have proven that they were right. Taking an approach similar to Vonk's and Beran's, neurobiologists Helen Ditz and Andreas

The more scientists learn about birds, the more they are realizing that the term "bird brain" is a compliment. Self-awareness, tool use, problem solving, observational learning, and theory of mind have been observed in birds. Now we can add numerical cognition to the list of bird talents, because research has demonstrated that birds can "count," too. Credit: Andreas Nieder, University of Tübingen, Germany

Nieder trained crows to peck at a computer touchscreen. Then they showed the crows two images containing different quantities (one to five) of dots.

The crows' task was to figure out if the two groups of dots contained the same number of dots. If they did, the crows were to peck the screen. If the quantities were different, the crows had to refrain from pecking. When they got the right answer, they were rewarded with treats.

The researchers made sure to include dots of different sizes and arrangements to ensure that the crows were responding to quantity rather than density or distribution. But this didn't stump the clever birds. It turns out that the old folk belief was based on truth. The crows were able to figure out whether the two groups of dots were the same or different—no matter the size or arrangement of the dots—and the only way they could have done this was by counting.

This experiment yielded another unexpected discovery: while the crows were busy counting, Ditz and Nieder observed their brain activity and discovered that certain neurons were consistently activated when the crows were thinking about numbers. Even more fascinating, those brain patterns were similar to human brain patterns when we think about numbers.

"ARITHMECHICKS"

Most people don't think of chickens when asked which animals they believe to be the most intelligent. But studies have demonstrated that chickens outperform dogs and even human toddlers in many cognitive and behavioral tests. As for chicks, within a few days of hatching, they can not only discriminate between different quantities, but also add and subtract.

Researchers Rosa Rugani, Laura Fontanari, Eleonora Simoni, and Lucia Regolin conducted a study that demonstrated chicks' amazing arithmetic abilities. They already knew that the chicks' "imprinting" behavior—the instinctive tendency to follow and remain physically close to their mother as soon as they hatch—could be projected onto inanimate objects they are consistently exposed to right after hatching. They also knew that the chicks tend to approach a group containing larger quantities

Within a few days of hatching, chicks appear to be capable of simple addition and subtraction.

Studies have shown that chickens outperform dogs and even human toddlers in many cognitive and behavioral tests!

of these imprinted objects. So as soon the chicks hatched, they put five small yellow balls in their nest so the chicks would imprint on them.

After a couple of days, the researchers tested the chicks. Placing each chick inside a glass enclosure, they moved the yellow balls (suspended from fishing line) behind one of two screens while the chick watched. They moved two balls behind one screen and three behind the other and then released the chick. All of the chicks immediately walked behind the screen that hid the larger quantity of balls.

Clearly the chicks needed a bigger challenge, so the researchers made the test more difficult. This time, they hid all the balls behind the two screens before bringing a chick into the enclosure. Then they moved the balls, one by one, back and forth, from one screen to the other, like a shell game. In order to keep track of how many balls were behind each of the two screens, the chicks had to keep track of the additions and subtractions of the balls as they appeared and disappeared behind the screens. When the chicks were released, they once again immediately went to the screen hiding the larger number of balls, suggesting that they can do basic arithmetic.

ALEX'S AMAZING MATHEMATICAL MIND

Alex was an African grey parrot who forever changed the way people think about animal intelligence. Irene Pepperberg, an animal cognition scientist, worked with Alex for more than thirty years. At the time of his death in 2007, Alex had learned more than one hundred words that he used correctly and creatively. Alex, whose name was an acronym for Avian Learning EXperiment, could identify actions, colors, shapes, materials, and quantities. He could count up to eight and make distinctions between attributes when counting. For example, when presented with a tray containing four green blocks, three red blocks, and five blue blocks, and asked "What color five?" Alex would respond "blue," suggesting he had distinguished between the three colors, determined the total number of blocks for each color, and answered correctly.

Pepperberg was led to explore Alex's abilities to do simple addition while testing another parrot named Griffin. She made two clicking sounds,

Alex could count and add together sets of objects. He even understood the abstract symbol for a quantity—the written Arabic numeral. For example, he could recognize the symbolic numeral "5" and correlate it to five objects! Credit: Arlene Levin for the Alex Foundation

hoping Griffin would vocalize "two." When Griffin remained silent, Pepperberg made two more clicks. This time, Alex, who was in the same room, spoke up, and said "four." Pepperberg continued with two more clicks and Alex said "six." Pepperberg soon realized that Alex could perform addition! By the end of his life, Alex could add up to three sets of objects as long as the total was eight or less.

Beyond counting and adding, Alex was able to understand and vocalize the abstract symbol for a quantity—the written Arabic numeral. It's one thing to see a group of objects and count them. It's another to see the purely symbolic numeral "5" and correlate it to five objects. Alex was first taught to identify plastic Arabic numerals that were placed on a tray, using the same vocal labels (words) that he had been taught to use to identify quantities of objects in a set. His trainers placed six Arabic numerals—each a different color—on his tray and trained him to respond to "What number blue?" or "What color six?" in the absence of any set of objects.

Once Alex was performing with approximately 80 percent accuracy, he was shown two plastic numbers of different colors and asked "What color number bigger?" or "What color number smaller?" The only way Alex could correctly answer that a green "2" was smaller than a blue "5" was if he understood the equivalence between the Arabic numeral and the quantity it represented. Amazingly, Alex figured this out on his own.

He even understood the concept of zero, which he had also seemed to discover on his own. He had been trained to respond "none" if none of the attributes in a set of objects were the same or if none were different. For example, if Alex was asked "What's same?" when shown a tray with two toys of different colors, shapes, and materials (such as a blue wooden triangle and a green plastic square), he would say "none;" and when asked "What's different?" when shown a tray with two toys that were identical (such as two blue wooden squares), he would also answer "none." Then, without any training, he applied the label, "none" to the absence of objects themselves, not just their attributes. For example, if asked "How many green?" about an object set with no green objects, he answered "none."

Sadly, Alex died in September 2007, shortly after Pepperberg had started working with him on these

Even the humble honeybee can "count"! Researchers designed a study in which bees flew into a maze through an entrance marked with either two or three dots. The bees had to remember the number of dots at the entrance in order to determine which path to choose when the maze forked into two paths, both of which were also marked with dots. The path marked with the same number of dots as the entrance led to a sugar water reward. For example, if the entrance was marked with three dots, the bees had to choose the three-dot path get to the reward. To be certain the bees weren't using other clues to find the sugar water, the researchers addressed scent, dot arrangement, and other variables. But the bees still chose the correct path, suggesting that they were able to "count," or distinguish between two dots and three dots.

Researchers at the Duke Lemur Center taught lemurs to use a touchscreen computer in order to find out if they can distinguish between quantities. Just as in the black bear and crow experiments, the lemurs had to choose either smaller or larger quantities of shapes and were rewarded with treats for choosing correctly. The lemurs performed well and were added to the ever-growing list of animals that can "count." Many scientists now believe that a sensitivity to quantity may be preprogrammed into the brains of a wide variety of animals, ranging from insects to primates.

Arabic numeral addition tests. Aside from the tremendous loss of a brilliant mind and charismatic companion, Pepperberg regretted the missed opportunity to show the world Alex's latest progress in arithmetic.

Bears, chickens, crows, and parrots aren't the only animals that can understand quantity or count. Ants, bees, chimps, dolphins, elephants, fish, gorillas, lemurs, monkeys, and spiders can count, too. It seems likely that having some kind of numeric cognition emerged a long time ago, early in evolution. Rather than being a skill that separates humans from other animals, a sensitivity to and awareness of number may be just another quality of mind that connects us to other species.

11. TECHNOLOGY ACROSS THE KINGDOM

TOOL USE

A CHIMPANZEE STONE AGE

Stone-Based Technology

In the opening scene of Stanley Kubrick's 1968 film *2001: A Space Odyssey*, an apelike protohuman discovers how to use a large bone as a weapon, communicating the idea of early humans' first use of a tool. After using the bone as a club to win a battle against a rival tribe, the protohuman triumphantly hurls the bone into the sky. While the bone is in midair, the film cuts to a satellite floating in space.

Kubrick's opening scene poetically represents a few million years of human evolution in a few seconds, and in doing so, reminds viewers of how far humankind has come as a species, thanks in large part to technology. And now, a recent archeological find is inspiring scientists to think about the potential influence of technology and evolution on another species. In 2006, researchers found evidence that chimpanzees from West Africa were using stones as tools to crack nuts 4,300 years ago. To put this timeframe in context, the human Stone Age ended about 6,000 years ago (it started roughly 2.6 million years ago). Unless much, much older chimpanzee tool

When primatologist Jane Goodall first observed chimpanzees strip the leaves off of twigs to fashion tools to fish for termites, it forever changed our definition of "human." Until then, scientists believed that only humans made and used tools. After Goodall's discovery, anthropologist and paleontologist Dr. Louis Leakey famously said, "Now we must redefine tool, redefine man, or accept chimpanzees as humans."

sites are discovered, this archeological find suggests that perhaps our primate cousins might be at the very beginning of their own chimpanzee Stone Age.

Although tool use in wild chimpanzee populations was first reported in the early nineteenth century, most scientists learned of chimpanzee tool use in 1960, when primatologist Jane Goodall was doing field work in Gombe National Park in Tanzania. At that time, Goodall observed chimpanzees strip the leaves off of twigs and then use the twigs as probes to "fish" for termites (a favorite chimpanzee snack). Goodall's observation was historic—it was the first time a scientist observed an animal make a tool and use it for a specific purpose. It was also the first time the news of tool-using animals was brought to the general public's attention through the relatively new medium of television.

Chimp Tech

In the fifty-plus years that have passed since then, we have learned that chimpanzees use an impressive number of tools. In fact, researchers now use the term "tool kits" to describe an array of tools regularly used by a group of chimpanzees. Most chimpanzee populations utilize about twenty different types of tools for a variety of functions, including hunting, gathering, personal hygiene, socializing, mating, and more. Certain chimpanzee communities even appear to share acquired technological knowledge with one another, supporting the idea of chimpanzee cultural transmission.

The tool kits of different chimp populations vary, but all chimps have certain tools in common. For example, all make leaf sponges to collect drinking water. Chimps crumple or chew leaves, dip the "sponge" in water, and then squeeze or suck the water into their mouths. According to Goodall, one can collect eight times as much water by dipping a leaf sponge into water as opposed to dipping one's fingers.

Although other animals use tools as well, one of the things that distinguishes chimps is their ability to use one type of material to make multiple kinds of tools. For example, in addition to using leaves as sponges, chimps use leaves as cups to transfer water to their mouth, as napkins to wipe materials from their bodies or other surfaces, and as temporary bandages or compresses that they hold against a wound.

What else can chimps do with plant material? They use long sticks as levers to move rocks and other heavy objects or to open and widen spaces. They use hefty sticks and small logs as clubs to break open beehives and in aggressive attacks on other animals,

Chimps use sticks for many purposes, including intimidation displays.

> The only other animal, besides chimps, that makes stabbing tools to kill other animals are humans.

including other chimps. They use smaller sticks as spades for digging, as puncture tools to extract honey, and as eating utensils to extract bone marrow and other edible parts from carcasses. They use leafy branches for many things: as umbrellas, back-scratchers, fly swatters, and material for bedding and pillows. Chimps also shake and wave branches in threat and intimidation displays.

Of course, the chimp's most famous use of plant material (discovered by Goodall) is their use of twigs, grasses, bark, and vines as probes to catch insects. Recently scientists observed chimps using plant material to catch a different kind of food—other primates. The only other animal that makes stabbing tools to kill other animals are humans. The unique population of chimps that were observed hunting "like people" lives in Senegal. These chimps fabricated spears by shaping the ends of branches into sharp points and used them to hunt small, nocturnal primates called galagos, also known as bushbabies, which often hide in hollow logs. The chimps would thrust their spears into the hollow logs and then withdraw them to look (or taste) for blood on the tips. If blood was present, the chimps continued to use the spears to capture the bushbaby, which they then consumed.

Chimps don't just use tools one at a time; they often use them in sets. By way of a human example, when we measure a length of lumber, saw it, and nail it together to build a bench, we are using a "tool set." It turns out that chimps have tool sets, too. One group of chimps in Gabon used a tool set consisting of five different tools just to collect honey—a "perforator" to puncture through the ground to locate the honey, a "pounder" to break open the hive, an "enlarger" to enlarge the hive's honey chambers, a "collector" to scoop the honey, and a "swabber" used as a spoon.

Also similar to humans, chimps use two-part, or "composite," tools. One of their most common composite tools is their version of the hammer and anvil. Stones or clubs are used as a hammer to crack

Chimpanzees have been using stones as tools for at least four thousand years. This male chimpanzee in Bossou, Guinea, is using two stones as a hammer and anvil to crack open nuts. Credit: Kathelijne Koops

open nuts or cleave large fruits on anvils made of stone or wood. Stones are also used as projectiles—along with other objects—during aggressive interactions.

Flintknapping Bonobos

One of the most fascinating examples of stone tool use is by two captive bonobos, Kanzi and Panbanisha. As part of a series of studies, researcher Sue Savage-Rumbaugh and her colleagues trained both Kanzi and Panbanisha in flintknapping, a technique to create sharp flakes of stone that was used by early humans to make some of the first stone tools, such as arrowheads and knives. At its simplest, knapping involves holding a stone core in one hand and using another stone as a hammer to strike off sharp pieces.

In the first study, both apes made flint flakes that they used to cut through rope and leather. In the second study, the apes made and used a wider variety of tools. Kanzi, in particular, showed a real talent for flintknapping and created tools that functioned as wedges, choppers, scrapers, and drills. He used these tools to access food that the researchers had hidden inside a log.

Give Them (Evolutionary) Time

If chimps are in their own Stone Age, then what kind of technology might they be capable of achieving given another few million years of evolution? Keep in mind that the human Stone Age lasted roughly 3.4 million years, so humankind had a very long technological infancy before finally inventing the smartphone. Perhaps, if given a chance, chimps might be capable of developing their own kinds of complex technology, just as humans have.

But chimps will likely never achieve world domination as portrayed in the science fiction film *Planet of the Apes* because another primate—humans—beat them to it. Chimpanzees have already disappeared from four African countries, and they are nearing extinction in many others due to habitat destruction and commercial hunting. Hopefully, with increased conservation efforts, chimps will be protected and have the chance to realize their own potential, in their own evolutionary time.

BAR TOOLS: THE CHIMPANZEE'S SHOT GLASS

About ten million years ago, a genetic mutation enabled both our human ancestors and apes to metabolize alcohol. About one hundred years ago, humans invented the shot glass. Not to be outdone, at some past but unknown point in time, chimpanzees

Humans and chimpanzees have a lot in common, including the ability to consume alcohol and to invent tools that make it easier to enjoy a drink.

invented the "shot leaf." Using their classic leaf sponge (chewed or crumpled leaves that are dipped into water and then squeezed to release the liquid into their mouths), chimps have been imbibing a lot more than water.

People living in Bossou, a village in the West African country of Guinea, tap raffia palm trees for the sap, which often ferments into alcohol. Over a seventeen-year period, researchers observed nearby chimps using their leafy drinking tool to do shots of fermented palm sap! The sap has an alcohol content that ranges from 6.2 to 13.8 proof, which is equivalent to beer and wine. According to researchers, the chimps drank heartily, averaging about a liter per visit to the tree taps. After their "happy hours," the intoxicated chimps found a cozy place to sleep off their buzz. No one knows if the chimps had hangovers the next day.

TOOLS OF THE NEW CALEDONIAN CROW'S TRADE

There are lots of species that *use* tools, but we only know of four species that *make* tools: humans, chimps, orangutans, and one of the Einsteins of the avian world, the New Caledonian crow. Not only do these brainy corvids belong to this elite group of toolmaking animals, they also are part of an even more exclusive class of toolmakers: animals that make hooked tools. To date, scientists have discovered only two species that make hooked tools: New Caledonian crows and humans.

Most people would guess that the most accomplished nonhuman toolmaker is the chimpanzee, but it's actually the New Caledonian crow. These intelligent birds not only fabricate complex tools but also improve them over time with design modifications. They even share their improvements and innovations with one another and pass them down to the next generation. Credit: Dr. Jennifer Holzhaider

To make their hooked hunting tool, which looks like a crochet hook, the crow meticulously prunes, trims, and shapes the end of a twig until it has a small hook. Once the tool is ready for action, the crow uses it to probe holes in rotted wood, insect nests, and other hard-to-reach places in search of larvae, worms, and other treats.

Not satisfied with just one truly impressive tool, New Caledonian crows developed a second distinctive tool: a serrated probe. Using his beak, the crow snips

Crows get attached to their tools, just like humans, stowing tools away in a hole or behind tree bark for later use.

and tears a narrow piece of leaf from a Pandanus tree—a palm-like plant with barbed leaves—until he has crafted what looks like a miniature hand saw. He uses this tool in two different ways: he either moves it quickly back and forth to broadly check for prey that might be hiding in leaf litter or other concealed areas, or he uses it to probe slowly, in a more deliberate motion, in an attempt to snag unsuspecting prey on one of the sharp barbs.

Just as carpenters often have a favorite hammer, these crows get attached to their tools, too. Scientists have observed them standing on their tools when not using them, and sometimes even tucking them into a hole or hiding them behind tree bark for later retrieval. By safeguarding their tools for future use, the crows do not need to spend time reinventing the wheel, or, in this case, the hook. This way, they have more time to hunt, eat, and possibly dream up another ingenious tool.

THE ROOK AND THE PITCHER

Inspired by one of Aesop's fables, British zoologist and Cambridge University professor Christopher Bird conducted a study on corvid intelligence. The tale "The Crow and the Pitcher" tells of a thirsty crow who finds a pitcher of water, but the water level is too low for him to reach. The clever crow drops pebbles into the pitcher until the water level rises and he can take a drink.

In Bird's experiment, he tempted four rooks (a corvid species that is similar to crows) with a worm that floated—out of reach—on the surface of water in a test tube. Bird videotaped the rooks as they appeared to assess the situation, circling the tube and considering their options. He then placed a pile of pebbles nearby. The rooks did exactly what Aesop's crow did: they turned the pebbles into displacement tools, filling the tube with them until the water level rose and they could get the worm. Even more impressive, the rooks selected the larger pebbles, which displaced more fluid than smaller pebbles, and so accomplished their goal in the fastest way. Bird's research showed that rooks' understanding of fluid mechanics is comparable to that of apes and children.

THE ANT'S TINY TOOLBOX

In the human world, agriculture often is regarded as marking the beginning of modern civilization. Humankind began farming about 11,500 years ago. But ants have us beat—they began farming about fifty million years ago.

Ants have been farming for about fifty million years longer than humans.

Like all ranchers, herder ants are protective of their herds. They fight off predators and protect the aphids' eggs, and when the eggs hatch into nymphs, they carry them to a food source to keep them happily "grazing."

> Herder ants keep other animals as livestock! They capture aphids, keep them in "herds," and "milk" them for a sugary substance to eat.

Leaf-cutter ants—the agriculturalists of the animal world—farm a tiny species of fungus called mycelium for food. To grow their crop, they use leaves as tools. First, they cut up leaves, which they carry back to their nests to chew into a pulp. Then they mix the leaf pulp with their own feces, which turns the mixture into a kind of manure. This fertilizer provides the nutrients for the fungus spores that they grow and store in their nests. Before long, the ants have a crop of "mushrooms" to feed the colony.

Herder ants engage in another agricultural practice: they use other animals as livestock. These little cowboy ants capture aphids, keep them in "herds," and use them in ways similar to the ways that humans use dairy cows—they "milk" the aphids for a sugary substance called "honeydew," which ants eat. Instead of making fences to keep the herd together, the ants excrete a chemical from their feet that has a tranquilizing effect on the aphids. Sometimes the ants even clip the aphids' wings so they cannot escape. Like ranchers, the ants also protect their herds from ladybugs and other predatory insects.

Beyond farming and herding, ants have a variety of other tools in their toolbox. To reduce competition for food, some species drop stones at the entrance of other ants' nests to blockade them and keep them from getting out and foraging. Other species use bits of leaves and soil as containers for carrying liquid food, such as honey and fruit pulp, which enable them to carry roughly ten times as much food as they can without containers.

Weaver ants—the fiber artists of the ant kingdom—make nests by connecting leaves and gluing the edges together with the sticky silk secreted by their larvae. The adults carry larvae in their jaws and very gently squeeze the larvae until they secrete a little silk at the point where the edges of two or more leaves meet. The ants move the larvae along the length of the leaf edges, squeezing them every step of the way like little tubes of construction glue, until the silk seams are complete.

12. FINDING THEIR WAY

SPATIAL INTELLIGENCE

SPYHOPPING SEALS NAVIGATE BY THE STARS

If you have ever looked out at the sea from the shore or a boat and noticed a little earless head suddenly pop out of the water, you probably saw a harbor seal. When a harbor seal wants to surface and check out the environment above water, he rises and "stands up" by rapidly moving his rear flippers back and forth in a posture called "spyhopping." Other marine mammals, such as whales and dolphins, also spyhop (but they do it by moving their tails). Marine mammals spyhop to search for prey or watch for predators, to get a better look at a tourist boat or kayaker, or to get their bearings by scoping out landmarks on the shore as reference points. And, according to researchers Björn Mauck, Nele Gläser, Wolfhard Schlosser, and Guido Dehnhardt, when a harbor seal spyhops at night, he might be engaging in a kind of celestial navigation similar to what ancient sailors used.

To understand how seals might be using the stars to navigate, you have to understand two astronavigation terms: *lodestar* and *azimuth*. A lodestar is a celestial object—such as the sun, the moon, a planet, or a single star—that is chosen as a point of reference for navigation. The azimuth is the point on the earth's horizon directly below the lodestar. For most of human seafaring history, sailors crossing the ocean plotted their direction by measuring the angle between the lodestar and the azimuth. This allowed them to estimate how far north or south they were from the equator. For example, the Vikings used Polaris—the North Star—as their lodestar, measuring the distance between it and the horizon to calculate their latitude. Mauck and his colleagues had a hunch that harbor seals might use a similar process to navigate.

To find out, the researchers turned a circular pool into a planetarium, raising a dome above it onto which they projected roughly six thousand points of light reproducing the stars and constellations of the Northern Hemisphere's night sky. They then separately trained two male harbor seals, Nick and

Malte, to locate a single lodestar—in this case Sirius, the brightest star in the Earth's night sky. They did this by highlighting Sirius with a laser pointer. The seals had to swim from the center of the pool toward the azimuth underneath Sirius. If they swam to the correct location, they were rewarded with food.

The seals were easy to train and were soon ready to take their astronavigational test. This time the researchers randomly spun the planetarium dome so the lodestar Sirius would be in a new, and unfamiliar, position to the seals, and they did not use the laser pointer. The seals were still able to locate Sirius and touch its correct azimuth point on the horizon.

Based on their ability to respond to star patterns, harbor seals might use astronavigation when traveling at night.

Humans are not the only animals that notice the stars. Seals, birds, and whales do, too, using stars as reference points to keep them on track during long migrations.

Early in the test, both Nick and Malte performed fairly well, but by the end they were finding the correct azimuth 100 percent of the time. Mauck and his colleagues noted the two seals' "outstanding precision" at identifying lodestars as one of the reasons they believe that seals can navigate by the stars. Of course, just because seals are capable of astronavigation doesn't mean they actually use this method to find their way in the dark. More research is needed before we can know that for certain. But this test does provide a plausible explanation for how seals navigate at night in the open sea.

Perhaps one of the most valuable aspects of this research is that it reminds us that humans are not the only animals that notice the stars. Beyond the evidence that seals may use stars for navigation, scientists have known for some time that certain birds can, too. They also speculate that whales might use stars as reference points to keep them on track on their long migrations. Humpback whales, for example, cover up to five thousand miles during their seasonal migration. They somehow manage to travel in nearly straight lines even when encountering currents, storms, and ever-changing sea depths. Scientists know that the humpbacks use the Earth's magnetic field to help keep them on course, but they doubt that magnetism alone is responsible for the accuracy of their beeline journeys. So it is possible that the whales, like harbor seals and birds, might be using the sun and stars to help them navigate.

There is something poetic and comforting in the realization that other species look up at the stars, just as we do. For all we know, birds, seals, whales, and other animals may see their own constellation patterns in the sparks of light that shine above us all.

THE SACRED SCARAB'S CELESTIAL MAP

In ancient Egypt, the scarab—a species of dung beetle—symbolized the lofty ideas of resurrection and self-generation. This may seem like powerful mojo to project onto modest little beetles, but their unique behavior helps explain its symbolism. To start with, the scarab encases her eggs in a ball of excrement and then rolls the sphere across the ground from east to west, which reminded the ancient Egyptians of the daily solar cycle in which the sun "dies" at sunset and is "resurrected" at sunrise. As the hidden eggs hatched, and young beetles emerged from the dung ball, it seemed to the Egyptians—who didn't know about the eggs—that the beetle possessed the power to self-generate. Believing that the scarab had special powers, they named it *kheper*, meaning "rising from, coming into being itself." This belief became enshrined in their pantheon as *Khepri*, a scarab-inspired god of rebirth and self-creation.

About five thousand years later, long after the cult of *Khepri* had faded into history, scientists discovered that the once-sacred beetle does have a heavenly connection, and "special powers," too. The story starts with a pile of animal excrement—dung beetle food and the material they use for making brooding balls (incubation chambers) for their eggs. Male beetles gather around the dung pile, take as much as they can carry (as much as fifty times their body weight), and shape it into marble-sized spheres. Each male then quickly rolls away his stash—which he will offer to a female as a brooding ball or food—but he has to move quickly to avoid getting robbed. It takes a lot of effort for a tiny beetle to make a marble-sized ball of dung, and some beetles take the easy road by stealing another's ball instead of making one of their own. The sooner the hard-working beetle sculpts his ball and gets away from the pile (where potential hijackers loiter), the more likely he gets to keep it. Navigationally speaking, the best escape strategy is to roll the dung ball away in a straight line (to avoid inadvertently circling back to the pile). And this is exactly what the dung beetles do, with rather uncanny precision.

Scientists wondered how dung beetles managed to chart such a precise course when rolling away their dung balls.

But before he rolls the dung ball away from the pile, each beetle performs a "dance" in which he climbs on top of his little sphere of dung and spins around. This isn't the only time the beetle does his dung dance: it also happens if the scarab encounters an obstacle or the ball rolls away. So, what is the dance on top of the ball all about?

A team of scientists from Sweden's Lund University were intrigued by the dung beetle's mysterious behavior, so they decided to investigate. What they figured out is that the beetle's dance is for orientation. When he climbs on top of his ball, the beetle checks the location of the sun or moon and uses that information to set a straight course.

Although that explained the dance, there was still a mystery that perplexed the scientists. On nights when the moon isn't visible, one would expect the beetles to have a tough time finding their way. How is it, then, that the beetles still manage to navigate in straight lines on moonless nights? The scientists' best guess was that the beetles were somehow navigating by the stars, despite their tiny eyes, which struck the scientists as too small to be able to reference pinpoints of distant starlight.

To find out whether the beetles were really navigating by the stars, the scientists set up an observation area in a planetarium so they could change the star patterns projected on the dome while observing the beetles' movements. They tested the beetles under a sky that included only the brightest stars, another sky that included just the soft glow of the Milky Way, and a complete sky that included all the stars and the Milky Way. They also tested the beetles without any view of the night sky at all, by putting little light-blocking hats on them, to see how they fared.

The beetles were fastest and followed the straightest paths under two conditions: when they had a view of the *complete* sky including the Milky Way, and when they had a view of the sky illuminated *only* by the Milky Way. Without being able to see the Milky Way, they were slow and moved in circuitous routes. They were also very disoriented when wearing the light-blocking hats. To see if the sensation of the hat was causing the beetles' disorientation, the scientists did the hat test again, using hats made of transparent material. While wearing those hats, the beetles behaved just as they did without any hats, suggesting that the issue was not with the hats but with their obscured view of the night sky.

Just give them a view of the Milky Way and dung beetles can chart a straight course.

While these tests showed that the beetles were indeed using celestial objects to navigate, and while it also appeared that the light from the Milky Way was particularly important to their method, the researchers still didn't understand how they did it. After all, the beetles were not continually climbing on top of their dung balls to double check the stars to make sure they were still on the right course, the way humans might keep checking their maps or GPS apps. To figure out this piece of the puzzle, the scientists went back to the planetarium and devised another experiment.

This time, they got creative with the planetarium's projections and rearranged the positions of the moon and stars. They then released the beetles and observed their response to this "alien" night skyscape. Contrary to expectation, the beetles did just fine. They danced, got their bearings under the new sky, charted a course, and made their characteristic beeline (or "beetleline") across the test area.

It seemed as if nothing could throw the beetles off course as long as they could see the sky. But then the scientists had another idea. This time they projected a new skyscape onto the dome *after* the beetles had done their orienting dance and started on their journey. The new celestial map did not match the one the beetles took their initial heading on, and that threw them off course. Finally, the scientists realized how the dung beetle accomplishes his characteristic straight-line getaways. He takes a mental "snapshot" of the sky while doing his dance on the dung ball, and then during his journey he compares this internal mental star map with the actual external starry sky to keep on track.

If only the ancient Egyptians could have known that when their sacred beetles were dancing on their tiny spheres of dung under the light of the firmament, they were making star maps in their minds.

CAN'T FIND YOUR CAR KEYS? TRY BEING A NUTCRACKER

As each summer comes to an end, the Clark's nutcracker—a member of the corvid family of birds that includes crows, ravens, magpies, and blue jays—prepares for winter by hiding stashes of food. This behavior, known as "food caching," is how birds and rodents store food in times of surplus for times of shortage. Each species has its own unique caching habits and strategies, but all caching requires remembering where food is stashed. The animal's survival depends on it.

Clark's nutcrackers are native to the mountainous regions of western North America, where they rely primarily on tiny pine seeds (about a centimeter long) as a food source. Using a special pouch under his tongue, the nutcracker can gather approximately ninety-five seeds at a time. Over the course of his caching period, a single nutcracker will collect about thirty thousand pine seeds and bury them in little batches in roughly five thousand different locations scattered over a territory that ranges from ten to hundreds of square miles. That behavior is pretty impressive in itself, but what is really impressive is the nutcracker's memory: he will remember—for up to nine months—all five thousand hiding spots. And he will do so even though the landscape changes throughout the seasons—as leaves fall, snow drifts, or new plants sprout.

Animals have developed all kinds of strategies for finding their way. Some use the sun, moon, and stars; others, such as the Saharan desert ant, count their steps. Mice use a technique that is more familiar to modern humans—they create signposts to help them remember their travel routes.

A single Clark's nutcracker caches tens of thousands of seeds and can find them as long as nine months later.

So, just how do they accomplish this feat of spatial memory? Scientists Stephen Vander Wall and Russell Balda believe that the nutcrackers create mental maps of significant landmarks and then mentally plot the location of their caches relative to these landmarks. They speculate that the nutcrackers might be using a triangulation technique, memorizing the distance and orientation from the landmarks to their cache. For example, a nutcracker might stash a cache near a rock, tree, and stream, and then remember the location by mentally noting that the cache is seven feet east of the stream, three feet south of the rock, and twelve feet north of the tree. (Of course, the nutcracker doesn't really measure in feet or compass degrees, but you get the idea.) Vander Wall tested this theory by going to a cache site and moving what he believed to be the nutcracker's landmarks. Sure enough, the nutcracker couldn't find his cache. And to think that we humans have a hard time remembering where we put our keys!

FOOD AND LODGING NEXT EXIT

Mice, which probably don't have the spatial memories of the nutcracker, may have figured out an easier and somewhat more "human" solution for finding their way: they use signposts. That is what Oxford University researchers David Macdonald and Pavel Stopka concluded after observing wood mice navigate while foraging. They noticed that as wood mice searched for food, they often left little objects, such as leaves, twigs, and shells, along their routes. To figure out what the mice were up to, the researchers set up a test area, removed the natural objects that the mice might use, and instead left them with a pile of little white discs. They then videotaped the mice.

The video showed the mice moving the white discs to locations that interested them and then orienting their foraging outings to and from the discs. The researchers determined that there was a statistically significant association between the white discs and the navigation routes of the mice. Their observations strongly suggest that mice, like humans, might be using markers as signposts to help them find their way around (next exit five miles) and also to help them—and other mice—locate areas of interest (food and lodging next exit).

13. FOR ART'S SAKE

CREATIVITY AND AESTHETICS

ANIMAL ART IN A HUMAN WORLD

Imagine a London art auction in 2005, during which a virtually unknown and deceased artist outsold French impressionist Pierre-August Renoir and pop artist Andy Warhol. Now imagine that this artist was a chimpanzee named Congo.

Born in 1954, Congo begin painting at age two when his caregiver, British zoologist Desmond Morris, put a paintbrush in his hand. The young chimpanzee took to painting immediately, and his career as an artist began. In his book *The Biology of Art*, Morris describes Congo's behavior while painting and drawing as focused and deliberate. Congo often asked to paint and disliked being interrupted while making art. He even screamed if somebody tried to take a painting away from him before he decided it was finished. It seemed to Morris that Congo found painting rewarding in itself, without any outside reinforcement.

The chimpanzee Congo created hundreds of works of art in the 1950s, most of which were exhibited and sold. Some of his work was purchased by such artistic luminaries as Pablo Picasso, Joan Miró, and Salvador Dalí. Painting by Congo

Gorillas Koko and Michael would title their art pieces, demonstrating what their art might represent. For example, Michael named a painting of a bouquet of pink flowers "Stink Pink More."

Left: Credit: Creative Commons BY-SA 2.0, Christine

Congo isn't the only captive animal known to create art. Other chimpanzees and orangutans, as well as gorillas, have also been taught to paint. Both Koko the gorilla and her deceased gorilla companion, Michael, painted. Sometimes they painted objects or companions (such as a dog); other times they painted from their imaginations. We know this because both gorillas could communicate through sign language and often titled their work, which allowed their caregivers to know what their art might represent. For example, Michael named a painting of a bouquet of pink flowers "Stink Pink More," and he titled a dynamic splatter of brown splotches tinged with black and green "Earthquake."

Orangutans seem to enjoy making art as much as other apes do, and they have worked in other media besides paint. Wattana, a female orangutan living at the Ménagerie du Jardin des Plantes (Botanical Garden Zoo) in Paris, prefers to work in fiber. She excels at knot-tying and uses her skills to create objects, adornments, and installations. Her caregivers never showed her how to make a knot, nor did they reward her for doing so. Wattana ties single, double, and triple knots, and combines them creatively—making multiple passes through loops—to create simple weavings and macramé-like forms that sometimes incorporate beads and other objects. She often ties the ends of strings to structures in her enclosure (such as ropes, poles, or wire mesh) and then uses these as a loom-like support for her knotting. Wattana can knot with her hands, feet, and mouth, and knots everything from string and shoelaces to toilet paper. She so enjoys knotting that sometimes, like a human artist caught up in the creative flow, she prefers to continue creating her art rather than eating.

> The average female orangutan builds roughly thirty thousand nests in her lifetime—which explains Wattana the orangutan's affinity for her knot-tying and weaving art pieces.

Knot-tying was considered a uniquely human activity until zoos and research centers started reporting that captive apes sometimes develop the ability to tie and untie knots. Because apes make nests in the wild, scientists believe that some apes build upon their instinctive nest-making skills when they learn to tie knots. Of all the great apes, orangutans in particular are especially good at knot-tying and weaving, no doubt because the females build elaborate nests for themselves and their offspring every night. In fact, the average female orangutan builds roughly thirty thousand nests in her lifetime. The nests—which they make by weaving together branches, sticks, and leaves—include a variety of amenities, such as mattresses, pillows, blankets, roofs, and sometimes even bunk beds. So if orangutans weave like this in the wild, it's not so surprising that a few have learned to knot in captivity. What *is* surprising is how Wattana's creative process is so humanlike. The more time she spends knotting, the better she gets at it, and this seems to result in an increased desire to knot. This is no different than a human learning to draw or play the piano—practice leads to proficiency, which brings pleasure and increases motivation.

Another member of the animal kingdom that has shown an interest in art-making is the dolphin. Animal trainer Frank Sanchez taught Chicky, a dolphin living in captivity in Switzerland, to scoot onto the deck surrounding her pool, take a paint brush (soaked with paint by Sanchez) in her mouth, and move the brush around a canvas. In between strokes, Sanchez reloaded the brush so Chicky could continue to paint. Sanchez usually rewarded Chicky with fish, but not always. Sometimes, Chicky wanted to paint even when fish were not offered as a reward. According to Sanchez, Chicky seemed to enjoy painting, and even occasionally paused to examine the canvas, as if considering her work in progress.

Dolphins demonstrate all kinds of creativity in captivity, ranging from theatrical ingenuity to problem solving. Wild dolphins are equally creative, using sea sponges as tools to dig for fish and protect their rostrums while foraging along the ocean floor.

Captive apes and dolphins aren't the only animal artists working with human art materials. Elephants have been trained to hold a paintbrush in their trunks and have created a wide variety of paintings. One elephant, Paya, even mastered representational art and is known for his elephant portraits. There are videos on YouTube that show Paya painting very strong likenesses of elephants—easily as good as a young child's art. But these astonishing representational portraits are achieved through conditioning by the trainers, who direct the elephants' strokes with subtle tugs on the elephants' ears or other cues not seen by the audience or cameras. Although the artworks are usually sold to help fund elephant conservation efforts, critics worry that the demand for elephant paintings might encourage more elephants to be taken into captivity in order to be trained to paint.

Many elephants have been taught to hold a paintbrush, but the paintings they create are usually the result of subtle directional cues provided by their trainers. Credit: Associated Press, Apichart Weerawong

How do most scientists assess these examples of captive animals making art? Most of them dismiss it for two primary reasons: first, all of these animals were given the tools for art-making by their human caregivers; and second, some of the animals are "coached" by trainers. Other scientists, however, believe that some—the apes in particular—demonstrate authentic creativity and even aesthetic sensibilities in their art-making. Regardless of which perspective one takes, it is clear that certain apes, such as Congo, Koko, Michael, and Wattana, not only developed an interest in art, but also derived pleasure from it and actively sought out the art-making process. In this regard, they are not so different from young children who fall in love with coloring when they are first given a box of crayons.

Mr. Bower Builds His Dream House

Wild animals exhibit creativity, too. Charles Darwin wrote that some birds have "fine powers of discrimination" and in some instances could be shown to "have a taste for the beautiful." Art is hard enough to define when it is created by humans, so what about when wild animals make objects that look like art? The best known animal artist in the wild—the bowerbird—builds creations that inspire scientists and philosophers to rethink the definition of art.

For millions of years, male bowerbirds have been building elaborate and visually striking "bowers" with twigs, leaves, nuts, flowers, fungi, moss, shells, insects, and other materials. Each bowerbird species has a unique template for building their bowers, but individuals improvise within the parameters of their species-specific design plan. Some species build spires; others build arbor-like structures, platforms, or avenues.

The satin bowerbird is the only other species—besides humans—known to kill and use another animal for the purpose of art.

These artistic bowers are built to impress females, with materials purposefully arranged in ways that

Each species of bower bird builds its own style of bower. The Satin bower bird builds a U-shaped bower that he decorates with blue objects, such as berries, feathers, flowers, and human debris.

female bowerbirds appear to find visually pleasing. Several species create beautiful gradients of color through the careful arrangement of objects. One species, the satin bowerbird, has an especially strong preference for the color blue and collects blue petals, shells, feathers, and human-made objects as building materials. If there is a shortage of blue material, he sometimes crushes fruit pulp to make blue "paint," which he then applies to his bower, occasionally using a piece of soft bark as a brush. These males are so obsessed with the color blue that they will even kill a bird in order to pluck its blue feathers to use as bower decorations. They also kill insects—such as beetles and butterflies—so they can use their wings and shells. Humans are the only other species known to kill and use other animals for art.

Male Satin bower birds are so obsessed with the color blue that they will even kill birds and insects so that they can use their blue feathers, wings, and shells to decorate their bowers.

Bowerbirds can take as long as a month to complete a bower. This is not surprising, as such a meticulous selection of materials in a particular composition takes time. They arrange and rearrange their materials as obsessively as any human artist. During their bower building, the males appear to step back, consider, and seemingly judge their own creations, deciding what looks right and what still needs work. This behavior has led some scientists and philosophers to surmise that they have an aesthetic sensibility.

On top of being a visual artist, the male bowerbird is a performing artist as well. When he has finished his masterpiece, he uses it as a theater and performs an impressive song and dance in front of or inside his bower, as a further demonstration of his suitability as a mate. If he impresses a female, he will get a chance to mate. But because most bowerbirds are not monogamous—they don't help the female build a nest (the bower is just for seduction), incubate eggs, or raise chicks—all the female gets out of it is his genes. But at least they are very likely to be artistic genes.

ALL THE WORLD'S A STAGE

Animal creativity isn't limited to visual art. Scientists are also discovering that some animals, both in captivity and in the wild, are accomplished performing artists. While living at the Language Research Center in Atlanta, two bonobos—Kanzi and Panbanisha—had the opportunity to jam with musician Peter Gabriel. The apes took turns "playing" a keyboard with Gabriel and his band. Gabriel described the apes as having a real sense of rhythm and musicality. Not to be outshined by the bonobos, in the summer of 2016, Koko the gorilla got to "play" bass guitar with the Red Hot Chili Peppers bassist, Flea.

Birds don't just sing to attract mates and defend territories. Brain chemistry analysis suggests that sometimes they sing for the same reason we do—because they enjoy it.

Apes may have revealed untapped musical potential in a studio setting, but when it comes to wild animals with musical talent, no group of animals comes to mind as quickly as birds do. In fact, we so strongly associate birds with music that the word we use for their vocalizations is *song*. For a long time, scientists believed that birds sing only to attract mates and defend territories. But musician and philosopher David Rothenberg had a different hypothesis. He believed that sometimes birds sing just for the pleasure of making music. Rothenberg formed his opinion, in part, by playing music with birds. As he played his clarinet, the birds sang along, responding in ways that suggested they were enjoying the "jam session." For years Rothenberg engaged in a spirited debate about bird music-making with some of the world's leading avian biologists, but he could not convince them that birds sometimes sing just because it feels good. Current research, however, supports Rothenberg. It turns out that when birds sing, their brains release dopamine, a pleasure hormone, which strongly suggests that sometimes they really do sing for the pleasure of it.

As for evaluating the artistry of birdsongs, we need to first think about what makes human music appealing. Most people would agree that music has the ability to evoke specific moods, such as tranquility or exultation, sadness or joy. Is the same true for songbirds? It appears to be, based on their behavioral responses to their own and other birds' songs. Some songs invite pair bonding and affiliation; others provoke aggressive behavior. Through acoustic analysis of recorded birdsongs, Rothenberg has come to believe that birdsong, like human music, creates a mood by using compositional techniques. The birds combine rhythms, pitches, and melodic transitions to create expectation, anticipation, tension, release, and surprise. So it turns out that some songbirds are composers.

Humpback whales—the great musicians of the sea—are composers, too, and they use a compositional technique that was once believed to belong exclusively to humans—rhyme. All the male humpbacks in a local group "sing" the same song, which changes from year to year. Since these songs are intricate, lengthy (some can last as long as twenty-two hours), and stylized compositions, how do all the whales learn them? To find out, biologists Linda Guinee and Katharine Payne analyzed more than five hundred humpback songs recorded over a twelve-year span and discovered that their songs contain organized phrases that rhyme. They speculate that this rhyming might serve the purpose of helping the whales learn a new song every year. Humans have an easier time learning and remembering information when it rhymes, so maybe whales do, too.

And beyond music, what about dance? Animals do that in their own way, too. Although males of many species perform impressive dances as part of courtship rituals, and some domesticated animals, such as dogs and parrots, have been trained to perform certain patterns of movement for fun or competitions, neither of these examples qualifies as the ability to dance—to find and keep a beat. For an animal to be judged as able to dance, he needs to demonstrate that he can spontaneously move with the beat, without training or demonstration by a human.

So what's the evidence that some animals can dance? One example is Snowball, a cockatoo who dances to the Backstreet Boys tune "Everybody." A YouTube video of Snowball strutting his stuff went viral on the Internet and caught the attention of neurobiologist Aniruddh Patel. To determine if Snowball was really moving in rhythm on his own, without training, Patel designed an experiment in which he altered the tempo of Snowball's favorite song. Using digital software, Patel created eleven different versions of "Everybody," ranging from 20 percent slower to 20 percent faster than the original. He then played each version to Snowball and videotaped the bird while he boogied to the various beats.

Although Snowball had far from a perfect score—in fact he was only on beat about 25 percent of the time—he was declared the first scientifically validated nonhuman dancer. Patel and other scientists

A sea lion was the first nonhuman mammal to demonstrate the ability to find and keep a beat.

believe that the probability of Snowball dancing on beat even just 25 percent of the time was too high to be dismissed as chance. It is more likely that Snowball was sometimes finding the beat and "dancing," and sometimes not. Snowball did much better when the song was played at its usual tempo, as one would expect. Try dancing to your favorite song sped up or slowed down, and it might take you time to find the beat, too.

There must be something about the Backstreet Boys because another animal, a sea lion named Ronan, dances to the same Backstreet Boys song. Ronan was the first nonhuman mammal to demonstrate the ability to find and keep a beat. Now scientists have discovered that elephants can dance, too. There are numerous videos on YouTube showing elephants keeping a beat, such as the video from the Pairi Daiza zoo in Belgium, which shows a group of elephants swaying their trunks and bodies in rhythm with classical music played by violinists standing a few

Art often raises interesting questions, especially when created by animals. This Celebes crested macaque picked up photographer David Slater's camera and took a selfie that became part of a big copyright debate. Slater claimed copyright ownership, but because only legal "persons" can create copyrightable work, and animals do not qualify for personhood, the U.S. Copyright Office refused to grant Slater the copyright. Although the idea of copyright ownership for animals may be interesting, the larger question of personhood for animals is far more important, raising questions about fundamental rights for nonhuman animals.

feet in front of them. Two elephants even swayed in unison. The zoo had arranged for the musicians to play for the elephants in order to get them used to live music before an upcoming concert that was to be performed at the zoo. They had no idea the elephants would react the way they did.

When it comes to animal music and dance, scientists are increasingly surprised at their abilities. But what about the other performing art—theater? Acting is linked to an instinct to imitate, and we know that many animals engage in mimicry and camouflage. But do animals mimic in creative ways? There are countless stories of parrots mimicking a dog's bark, doorbells, and other sounds in ways that often seem creative, even if only as a means to a mischievous end. Dolphins are another animal with a talent for mimicry, having demonstrated spontaneous, untrained imitation that borders on theater. In his book *Beyond Words: What Animals Think and Feel*, Carl Safina recounts a few stories about captive dolphins that suggest they are capable of theatrical creativity.

The first story is about a dolphin named Daan. When human divers cleaned algae from the viewing ports of his tank he would sometimes watch them. One day, after watching the divers, he picked up a seagull feather that had fallen into the tank and began stroking the viewing port glass, copying the movements of the divers. Daan even mimicked the human divers' habit of steadying themselves by holding onto the bars next to the window—Daan would position one of his flippers in the same spot.

Another dolphin named Haig mimicked the motions used to vacuum the tank. Divers used an underwater vacuum to clean the bottom of the tank and sometimes left the vacuum in the tank overnight. One morning Haig picked up the vacuum between her two pectoral fins and pushed it across the bottom of the tank.

At that same aquarium, a young calf named Dolly was curious about the visitors who gathered to observe her. She must have watched them carefully, because after one of the visitors who was smoking a cigarette exhaled the smoke toward the viewing port, Dolly swam to her mother, nursed, and returned with a mouthful of milk that she released in front of the window. This may be the first recorded example of a dolphin producing a theatrical "special effect."

14. RE-IMAGINING IQ

THINKING OUTSIDE THE HUMAN BRAIN

THE MOLLUSK WITH A MIND

An Alien Among Us

Octopus researchers often say that if we want to imagine what extraterrestrial intelligence might look like, we need look no further than the cephalopods. About five hundred million years ago, cephalopods (octopus, nautilus, squid, and cuttlefish) evolved from a simple snail-like ancestor. Over time, these squishy, soft-bodied creatures became what some scientists have described as the first intelligent life on Earth.

With eight limbs that attach to his head and three hearts that pump blue blood, the octopus's anatomy is undeniably otherworldly. His boneless body can squeeze through any opening larger than his beak, which, by the way, is located on the underside of his body where his arms meet. Inside his parrot-like mouth is a toothed tongue and flesh-dissolving neurotoxin saliva. Octopus eyes, which are the most complex of any invertebrate, can rotate in any direction and perceive polarized light.

Each octopus arm has roughly two hundred suckers, which can fold into grasping positions similar to the way the human thumb and forefinger come together. Each sucker has about ten thousand receptor cells for touch, taste, and smell, which makes octopuses about a thousand times more sensitive to those stimuli than humans.

The octopus can change the color, pattern, and texture of his skin to communicate mood and readiness to mate, blend into his surroundings, or mimic

> Using their dexterous arms, octopuses can climb out of the water, walk on land (and even "run"), pass balls from sucker to sucker, open childproof jars, dismantle LEGO® constructions, untie knots, and escape from their enclosures!

Octopuses can do amazing things, from changing their skin color, pattern, and texture to blend into their environment, to navigating mazes and using tools.

another species. Another trick up his eight sleeves is his ability to disappear in a cloud of ink that he blasts out of his siphon—a tube-like structure through which water, ink, and bodily waste flows. He also uses his siphon to swim (shooting water out of it in a kind of jet propulsion technique), move objects, uncover prey buried in sand, clean his den, and blast unwanted annoyances.

If an octopus loses an arm, no problem—he can regenerate it, which is convenient because he has learned to make the most of his many arms. When using his arms as "legs," this spidery cephalopod can climb out of the water, walk on land, and even "run," sometimes moving as quickly as a cat. Using his arms like hands, the octopus demonstrates impressive dexterity: he can pass a ball from sucker to sucker, open clams and childproof pill jars, dismantle LEGO® constructions, build stone walls, untie knots (one octopus even removed his own surgical sutures), and escape from all kinds of enclosures.

There are, in fact, countless stories of octopuses escaping their tanks in labs and aquaria. Sometimes they crawl across the floor to another tank, slip inside, and snack on unsuspecting fish and crustaceans. Other times they escape back into the sea, leaving only suction-cup tracks as clues. This was what happened in 2016 when Inky, an octopus at New Zealand's National Aquarium, broke out of his tank and slipped back into the Pacific through a drainpipe.

In *Kraken: The Curious, Exciting, and Slightly Disturbing Science of Squid*, Wendy Williams recounts the story of Lucky Sucker, a two-spot octopus found "walking" down a sidewalk in Long Beach, California. Fortunately, a student rescued her and brought her to the nearby Aquarium of the Pacific, where she lived out her days. No one knows for certain how Lucky found herself in downtown Long Beach. James Wood, a cephalopod scientist who worked at the aquarium at the time, speculates that she might have been on a refrigerated fish truck as by-catch—the unwanted creatures caught by mistake during commercial fishing—and escaped. Regardless of what twist of fate brought Lucky to the aquarium, Williams described her as the "Greta Garbo" of the cephalopod world. According to Williams, Lucky had star appeal and really knew how to engage an audience by "strutting her stuff" like "an actress in her screen debut."

More than fifty years before Inky and Lucky became famous for their adventures, Jacques Cousteau wrote about an errant octopus in his 1973 book *Octopus and Squid: The Soft Intelligence*. According to Cousteau, a friend of his placed an octopus in an aquarium and covered it with a weighted lid to prevent his escape. Not long afterward, his friend found the octopus in his library, exploring books and turning the pages with his arms.

The Smartest Mollusk in the Sea

Regarded as the most intelligent invertebrate on Earth, the octopus is also a tool user. Several species seem to carefully choose stones, shells, bits of glass, and other objects to stack at the entrance to their dens, presumably to create smaller openings less easily penetrated by predators. Some species have been observed using stones as wedges to prop open the shells of bivalves while they reach inside to access the flesh. One species even builds "mobile homes." The veined octopus collects discarded coconut shells and assembles them into a makeshift clam-like shelter for "camping" away from home.

The octopus's arms have "minds" of their own—roughly half of an octopus's neurons are not in her brain, but in her eight arms.

Perhaps the most interesting example of octopus tool use is by a group of octopuses known as the blanket octopuses, named after the long transparent arm webs of adult females. Blankets are immune to the venom of the Portuguese man o' war jellyfish, so when they run into one, they sometimes help themselves to a venomous tentacle by breaking it off. Then, like the Zorros of the marine world, with their webs trailing behind them like a dashing cloak, they use the toxic jellyfish tentacles like swords to ward off predators!

In research labs, octopuses have solved simple mazes in order to access food or desirable dens. They have learned how to do this on their own, and on at least one occasion, an octopus learned by watching another octopus, which suggests they might be observational learners. Octopuses have a playful side, too, and often engage in play-like behavior with toys and other objects placed in their tanks.

As you might expect with such a multifaceted animal, octopuses appear to have personality. Most scientists—as well as people who keep pet octopuses—insist that each of these charismatic cephalopods is his own "person." Octopus researcher Jennifer Mather developed a personality test for octopuses in which individuals are categorized by their responses to different kinds of stimuli and then compared to one another. Mather concluded that octopuses have specific temperamental traits and unique personalities, with preferences and dislikes for everything from food and toys to people. When it comes to their preferences for people, octopuses not only tell people apart by taste and smell, but studies show that they can even recognize human faces.

Brainy Bodies

Although all of these octopus intelligence traits and talents are impressive, what makes octopuses really stand out is how they think. Octopuses have neurological systems that are very, very different from ours. The octopus's brain—the largest brain of any invertebrate—is wrapped around his esophagus and contains roughly 400 million neurons. This number, in itself, is not so impressive. It's more neurons than a guinea pig has (240 million), but much less than a cat has (760 million), and significantly less than a human has (86 billion). What is impressive is that roughly half of an octopus's neurons are located in his eight arms. Having neurons distributed throughout his arms means the octopus has a decentralized intelligence. This is why when an octopus loses an arm, for hours afterwards the severed arm can still crawl around and pick up objects.

But these arms with minds of their own don't just have motor control skills—they can "see," too. A study by University of California–Santa Barbara scientists Desmond Ramirez and Todd Oakley discovered that the skin of the California two-spot octopus is capable of sensing light without input from the central nervous system. The octopus's skin—which contains the same light-sensitive proteins found in his eyes—is able to detect light and discern brightness levels, which are then used to determine a camouflage strategy.

Once an octopus decides how he's going to camouflage himself, his three layers of skin and muscles take over. The outer layer of skin has cells called chromatophores, which are pigment sacs surrounded by a ring of muscles. These cells create colors such as yellow, orange, red, and black. When the octopus flexes his chromatophore muscles to expand the sac, the pigment is revealed. When he contracts the muscles, the pigment is concealed. The middle layer of skin has iridophores, which are cells that produce blues and greens by reflecting the colors of the environment. At the deepest layer are the leucophores, cells that produce white light and act as a canvas for the other colors of the octopus's light show. The muscles receive signals from the octopus's brain—as well as his arms—and can change the octopus's colors and patterns in less than one hundred milliseconds.

And then there's texture. The octopus uses a different set of muscles to raise his skin into peaks or

A mimic octopus can imitate a banded sea snake slithering across the sea floor.

flatten it into smoothness. He can make himself look like a rock, coral, seaweed, or other part of his environment. But his camouflaging isn't just used to hide from predators; it is also used for hunting. The octopus sometimes uses a display known as "passing cloud" in which he creates an "animation" of a shadow-like shape passing over his body. He uses this illusion of movement to scare prey out of their hiding places so he can pounce on them. By not actually moving his body but only creating the appearance of movement, the octopus has a better vantage point for watching and then catching his prey.

Some species of octopus use their brainy bodies to morph into the appearance of other animals. Scientists have observed the mimic octopus using his camouflaging abilities to avoid predators by impersonating lion fish, sea snakes, and sole (a kind of flat fish)—all of which are highly poisonous. Divers also have reported seeing mimics impersonating anemones, jellyfish, mantis shrimp, feather stars, brittle stars, giant crabs, seahorses, crocodile snake eels, stingrays, and nudibranchs.

Amazingly, the mimic will not only alter his shape, color, pattern, and texture, but even his behavior. And even more impressive than his mimicry skills is his knowledge of marine ecology. The mimic octopus seems to know which deadly creature he should impersonate based on local who-eats-who ecology. For example, when scientists observed a mimic being attacked by damselfishes, they saw it imitate a banded sea snake, a known predator of damselfishes.

chimp memory champ

Each of the two competitors watched a computer screen that flashed five numbers between one and nine. The numbers appeared for less than a fifth of a second at different locations on the screen and then vanished behind white squares. The challenge was to remember the position of the numbers and then touch the squares that concealed the numbers, in ascending numerical order.

Scientists believe that chimpanzees have a naturally photographic memory, which probably arose out of an evolutionary need to keep track of other chimps and food sources.

The older competitor, Ben Pridmore, was a thirty-year-old accountant from the United Kingdom and a three-time World Memory Champion who held the record for memorizing—in less than thirty seconds—the order of a shuffled fifty-two-card deck. The younger competitor, Ayuma, was a seven-year-old chimp raised in captivity in Japan. The results: Pridmore scored correctly 33 percent of the time. Ayuma scored correctly about 90 percent of the time, performing almost three times as well as the human World Memory Champion!

Ayuma has an amazing memory compared to humans, and he is special among chimps, too—but not that special. Most chimps can beat most humans at short-term memory games. Japanese researcher Tetsuro Matsuzawa and other researchers believe that Ayuma's performance is the result of chimpanzees' natural photographic memory, which probably arose out of an evolutionary need to keep track of other chimps and food sources. Ayuma's performance may also help foster humility in humans.

CATCHING FISH WITH LOGIC

If most people heard that an animal performed better than an average ten-year-old child at abstract problem solving, they would probably imagine the animal to be an ape or dolphin. But the animal with this distinction is a captive sea lion named Rio. She and other sea lions have successfully completed cognitive tasks that many primates—and even some children—have been unable to.

Biologist Colleen Reichmuth taught Rio to associate certain sounds with either a number or letter (assigned randomly). For example, the sound of chirping crickets was assigned to the letter "D" and the sound of a ringing phone was assigned to the letter "B." So when Rio hears the chirping crickets and is shown two choices—the letter "D" and another letter—she has to point her snout at the correct answer—"D" —if she wants her trainers to toss her a fish.

Once Rio figured out this first test, Reichmuth taught Rio that all the letters of the alphabet belong to one group (letters), and all numbers belong to another group (numbers). She did this so that she could see if Rio could use logic to make a choice. Once Rio learned to which group numbers and letters belonged, Reichmuth set up an experiment that required Rio to make a logical deduction.

Why would a sea lion be able to think logically? Scientists believe that the ability to draw logical conclusions may have evolved to help animals solve complex social problems in the wild, such as figuring out who is fun to hang out with versus who is most likely to start a fight.

Reichmuth played the sound of the ringing phone and showed Rio the letter "C" and number 9. Neither was the correct answer—remember that Rio had previously learned that a ringing phone was associated with the letter "B." However, "C" belongs to the same group (letters) as the correct answer "B," so it is a more correct answer than 9 is. Rio quickly figured out that "C" was the correct answer, and received her fish reward. It appeared that she was able to deduce that without the option of "B," the next best choice was one that belonged to the same group as "B," which was letters. Amazingly, Rio chose the correct answer in these kinds of tests about 90 percent of the time, proving that she was able to think logically. She performed feats of logic never observed in any other captive animal, and earned a lot of fish rewards in the process.

The Art of a Dolphin's Deal

At the Institute for Marine Mammal Studies in Mississippi, dolphins were trained to keep their pools clean by collecting litter, which they could trade for fish. The pool stayed clean and the dolphins seemed pleased to get a fish for each piece of trash they collected. Then one of the dolphins, named Kelly, had an idea . . .

One day Kelly found a piece of paper trash, took it to the bottom of the pool, and tucked it under a rock. Later, when she saw a trainer, she swam down to her trash stash, tore off a tiny piece of paper, and presented it to the trainer for a fish. Kelly figured out that by turning a single piece of trash into many pieces of trash, she could maximize her rewards.

Kelly's creative strategy suggested a lot about her level of intelligence. It showed that she was capable of delayed gratification, which requires an awareness of and planning for the future. It may also indicate that she had an understanding of the trainers' mental states, because it seemed like she was hiding the piece of trash, anticipating the trainers' possible disapproval. By portioning the paper, Kelly also showed a shrewd understanding of the rules of the trash-trading deal and how to take advantage of them. She was able to make what appears to be a logical deduction: "one piece of trash, one fish; turn that one piece of trash into many pieces of trash, get more fish." She was optimizing her resources to increase her rewards.

It wasn't long before Kelly demonstrated even more creativity in her bid for more fish. One day, she caught a gull that was scavenging around the pool. When she presented it to the trainers, they rewarded her with not just one, but a whole lot of fish. That bounty must have set off a light bulb in Kelly's head, because that night when she was given her supper, she saved one of the fish, tucking it under her

Dolphin studies have showed that these charismatic marine mammals possess impressive critical and creative thinking skills.

trash-stash rock. When no one was watching, she used the fish to bait another seagull, caught it, and later presented it to the trainers. Once again, Kelly was rewarded with a bounty of fish.

Clearly, Kelly understood that the seagulls were worth more fish than pieces of trash were. Thereafter, she regularly saved fish from her dinner or previous rewards as bait for seagulls. And that's not all. She soon taught her calf to do the same—and her calf eventually taught other dolphins.

THE SEARCH FOR INTELLIGENT LIFE

Once, when discussing the human bias in scientists' approaches to studying the intelligence of other species, Dr. James Wood, a cephalopod researcher, offered an amusing tongue-in-cheek analogy: "Imagine if an octopus made an intelligence test for humans. It might have a question that goes, 'How many color patterns can your severed arm produce in one second?'" If other species were to collaborate on a hypothetical IQ test for humans, the test would become even more challenging. Clark's nutcrackers might ask us to hide five thousand seeds and see how many we could find weeks later; a dung beetle might propose we take a mental snapshot of the night sky and use that star map in our minds to navigate; and chickens might suggest we compare the arithmetic abilities of week-old babies to week-old chicks.

Most humans would perform very poorly on a test with challenges like these, which supports Wood's point about how human bias influences the way intelligence is defined and studied. Due to cultural differences, it's hard enough designing intelligence tests for people, let alone animals. As scientists learn more about the minds of animals, they are realizing that every species has its own kind of intelligence. To better understand these different intelligences,

The impressive capacities of octopuses—and of many other animals—force us to reconsider how we define and study intelligence in other species.

scientists have come to realize that each species needs to be approached on its own terms.

When scientists consider where a species lives; how it senses, perceives, and engages its world; and how its intelligence might have evolved in response to these factors, they discover that species are more intelligent than they initially appear. For example, for aquariums with an octopus, a favorite aquarium demonstration for visitors involves placing the resident octopus in a large glass jar and screwing on the lid. Audiences watch in amazement as the notoriously dexterous octopus unscrews the lid from the inside and escapes.

So when octopus researchers put a crayfish—a favorite octopus food—inside a lidded glass jar and put it in a tank with an octopus, they were surprised to see the octopus just sit there, uninterested in opening the jar to get a free meal. At first, researchers wondered if they had overestimated the intelligence of the octopus. But eventually they figured out that because the octopus uses touch and smell to find prey in its natural habitat, the sealed crayfish was essentially undetectable. The researchers smeared fish scent on the outside of the jar and tried again. This time the octopus went right to work—he swiftly unscrewed the lidded jar and ate the crayfish.

This kind of approach—in which scientists consider the test from the animal's point of view—is relatively recent. Too often in the past, scientists drew faulty conclusions about animal intelligence because they didn't ask the right questions. Consequently, many animals were misunderstood and judged to be less intelligent than they actually are.

Theoretical physicist Werner Heisenberg once said, "What we observe is not nature itself, but nature exposed to our method of questioning." As scientists take human bias into account, they are asking more creative questions, and in doing so, they are revealing many kinds of unimaginable and remarkable intelligences and are finally answering the question, "Are we the only intelligent life in the universe?"

The answer is clearer than ever: we are not.

notes

introduction

xiii **One day we humans:** Excerpt from the film, *The Whale*, as quoted in a pdf posted on www.bullfrogfilms.com, http://www.bullfrogfilms.com/guides/whaleguide.pdf.

xiii **Shooter, the resident elk in the Pocatello Zoo:** "Caught on Camera: Lifeguard Elk on Duty," YouTube video, 2:18, from a report televised by KPVI News 6 and posted online on June 23, 2011, https://www.youtube.com/watch?v=KJWIMnBW9xQ; Vanessa Grieve, "Elk saves marmot," *Idaho State Journal*, July 2, 2011, http://idahostatejournal.com/news/online/elk-saves-marmot/article_3dc68360-a504-11e0-b25a-001cc4c03286.html.

xiii **That's what one of the zookeepers, Dr. Joy:** Krishna Strong (zookeeper, Pocatella Zoo, Idaho) in discussion with the author, December 2016.

xiv **Primatologist Frans de Waal suggests:** Frans B.M. de Waal, *Are We Smart Enough to Know How Smart Animals Are?* (New York: W. W. Norton & Company, 2016), Kindle edition, location 4134.

xv **The great American naturalist:** Henry Beston, *The Outermost House: A Year of Life on the Great Beach of Cape Cod* (New York: Henry Holt, 1992), 25.

chapter 1: the joke's on us: laughter, humor, and mischief

3 **Best place to tickle a rat:** Jeff Burgdorf and Jaak Panksepp, "Laughing rats and the evolutionary antecedents of human joy?," *Physiology & Behavior* 79 (2003), 533–47, doi:10.1016/S0031–9384(03)00159–8.

4 **Panksepp believes that the:** Jaak Panksepp, "Beyond a Joke: From Animal Laughter to Human Joy?" *Science*, 308, no. 5718 (April 2005): 62–63, doi:10.1126/science.1112066.

5 **Georgia, a captive chimp:** Frans B.M. de Waal, *The Bonobo and the Atheist: In Search of Humanism among the Primates* (New York: W. W. Norton & Company, 2013), 125.

6 **Once, to tease one of her teachers:** Steven Wise, *Drawing the Line: Science and the Case for Animal Rights* (Cambridge, MA: Perseus Books, 2002), 219.

6 **Instead, her trainer, Penny Patterson:** Francine Patterson and Wendy Gordon, "The Case

for the Personhood of Gorillas," in *The Great Ape Project*, eds. Paola Cavalieri and Peter Singer (New York: St. Martin's Griffin, 1993), 66–67.

6 **Dolphins may not share:** Jonathan P. Balcombe, *Pleasurable Kingdom: Animals and the Nature of Feeling Good* (London: Macmillan, 2006), 84.

6 **Parrots are notorious pranksters:** Eugene Linden, *The Parrot's Lament and Other True Tales of Animal Intrigue, Intelligence, and Ingenuity* (New York: Dutton, 1999), Kindle edition, 40.

7 **According to Poulsen, bears laugh:** Else Poulsen, "Bears aren't just smart—they're funny too," AnimalsAsia "Media News," October 24, 2014, www.animalsasia.org, https://www.animalsasia.org/us/media/news/news-archive/bears-aren%E2%80%99t-just-smart-%E2%80%93-they%E2%80%99re-funny-too.html Else Poulsen and Stephen Herrero, *Smiling Bears: A Zookeeper Explores the Behaviour and Emotional Life of Bears* (Vancouver, BC: Greystone Books, 2009).

9 **Dogs pant in a specific way:** Patricia Simonet, Donna Versteeg, and Dan Storie, "Dog-laughter: Recorded playback reduces stress related behavior in shelter dogs," originally from *Proceedings of the 7th International Conference on Environmental Enrichment* (July 31–August 5, 2005), posted on www.Pettalk.org, http://www.petalk.org/LaughingDog.pdf.

Chapter 2: A Generous Nature: Reciprocity and Cooperation

11 **Derek was a crow:** Wildlife rehabilitator (anonymity requested) in discussion with the author, September 2014.

12 **In 2015, one such story:** Katy Sewall, "The girl who gets gifts from birds," *BBC News Magazine* online, February 24, 2015, www.bbc.com, http://www.bbc.com/news/magazine-31604026.

12 **An ornithologist discovered and freed:** John Marzluff and Tony Angell, *Gifts of the Crow: How Perception, Emotion, and Thought Allow Smart Birds to Behave Like Humans* (New York: Free Press, 2012), 113.

13 **Paul Nicklen was swimming:** Paul Nicklen, *Polar Obsession* (Washington, D.C.: National Geographic, 2009), 156–58.

14 **Cooperation in capuchin monkeys:** Frans B.M. de Waal, "How Animals Do Business," *Scientific American* (April 2005): 72–79.

14 **In a similar experiment:** Joshua M. Plotnik, Richard Lair, Wirot Suphachoksahakun, et al., "Elephants know when they need a helping trunk in a cooperative task," *Proceedings of the National Academy of Sciences* 108, no. 12 (March 2011): 5116–21, doi:10.1073/pnas.1101765108.

15 **In the coastal city of Laguna:** Karen Pryor and Jon Lindbergh, "A Dolphin-Human Fishing Cooperative in Brazil," *Marine Mammal Science*, 6, no. 1 (1990): 77–82, doi:10.1111/j.1748–7692.1990.tb00228.x; Joe Roman, "Fishing With Dolphins," *Slate* "Science," January 31, 2013, www.slate.com, http://www.slate.com/articles/health_and_science/science/2013/01/fishing_with_dolphins_symbiosis_between_humans_and_marine_mammals_to_catch.html.

16 **Dolphins are not the only aquatic animals:** Simon J. Brandl and David R. Bellwood, "Coordinated vigilance provides evidence for direct reciprocity in coral reef fishes," *Scientific Reports* 5, no. 14556 (September 2015), http://www.nature.com/articles/srep14556, doi: 10.1038/srep14556.

Chapter 3. Fair and Square: Playing by the Rules

19 **Watch the YouTube video:** "Two Monkeys Were Paid Unequally," YouTube video, 2:38, posted by "addy," December 18, 2012, https://www.youtube.com/watch?v=_Go8tnl21MU, excerpted from the filmed TED Talk "Moral behavior in animals," by Frans de Waal, www.Ted.com, posted April 2012, https://www.ted.com/talks/frans_de_waal_do_animals_have_morals.

19 **The idea for the now famous experiment:** Sarah F. Brosnan and Frans B.M de Waal, "Monkeys reject unequal pay," *Nature* 425 (September 18, 2003): 297–99.

20 **Brosnan set up a similar experiment:** Sarah F. Brosnan, Catherine Talbot, Megan Ahlgren, et al., "Mechanisms Underlying Responses to Inequitable Outcomes in Chimpanzees, Pan Troglodytes," *Animal Behaviour* 79, no. 6 (June 2010): 1229–37; PDF version posted at ScholarWorks @ Georgia State University, Department of Psychology, Georgia State University, https://pdfs.semanticscholar.org/0e97/6448ca88c905f142c-d3a2c1444685efac059.pdf, doi:10.1016/j.anbehav.2010.02.019.

21 **They were put through a similar test:** Claudia A. F. Wascher and Thomas Bugnyar, "Behavioral Responses to Inequity in Reward Distribution and Working Effort in Crows and Ravens," *PLoS One* 8(2), (February 20, 2013): e56885, doi:10.1371/journal.pone.0056885, http://journals.plos.org/plosone/article?id=10.1371/journal.pone.0056885.

21 **Exploring still other species:** Friederike Range, Lisa Horn, Zsófia Viranyi, et al., "The absence of reward induces inequity aversion in dogs," *Proceedings of the National Academy of Sciences* 106, no. 1 (January 6, 2009): 340–45, http://www.pnas.org/content/106/1/340.full.pdf, doi.org/10.1073/pnas.0810957105.

22 **Dogs don't just want to be fairly rewarded:** Marc Bekoff, "Playful fun in dogs," *Current Biology* 25, no. 1 (January 2015): R4–R7; Marc Bekoff, *Animals at Play: Rules of the Game* (Philadelphia: Temple University Press, 2008); Marc Bekoff and Jessica Pierce, *Wild Justice: The Moral*

Lives of Animals (Chicago: University of Chicago Press, 2009).

23 **Hanging out with a cheater might be okay:** Jorg J. M. Massen, Caroline Ritter, and Thomas Bugnyar, "Tolerance and reward equity predict cooperation in ravens (*Corvus corax*)," *Scientific Reports* 5 (October 7, 2015): 15021, doi:10.1038/srep15021, http://www.nature.com/articles/srep15021.

23 **When it comes to cheating:** Stephen J. Dubner and Steven D. Levitt, "Monkey Business," *New York Times Magazine* online, June 5, 2005, www.nytimes.com, http://www.nytimes.com/2005/06/05/magazine/monkey-business.html; Mark Buchanan, "Money and Monkey Business," *The New Scientist* 2524 (November 5, 2005): 40–43; PDF version posted at http://www.anderson.ucla.edu/faculty/keith.chen/articles/NewScientist%20text%2011_5_05.pdf.

Chapter 4. Stand by Me: Friendship

27 **A primatologist once joked:** Joan B. Silk, "Using the 'F'-word in Primatology," *Behaviour* 139, no. 2 (February 2002): 421–46; PDF version posted at http://www.sscnet.ucla.edu/anthro/faculty/silk/PDF%20Files%20Pubs/F-word.pdf.

27 **Even rattlesnakes may have friends:** Amanda Pachniewska, "Snake Researcher Melissa Amarello," *Animal Cognition*, January 12, 2016, www.animalcognition.org, http://www.animalcognition.org/2016/01/12/interview-with-snake-behavior-researcher-melissa-amarello/.

27 **One of the hallmarks of friendship:** "The Hidden Lives of Ducks and Geese," *PETA* "Issues," www.peta.org, http://www.peta.org/issues/animals-used-for-food/factory-farming/ducks-geese/hidden-lives-ducks-geese/.

27 **Compassion is another aspect of friendship:** Joshua M. Plotnik and Frans B. M. de Waal, "Asian elephants (*Elephas maximus*) reassure others in distress," *Peer J* 2: e278 (February 18, 2014), doi:10.7717/peerj.278, www.peerj.com, https://peerj.com/articles/278/.

28 **Shirley and Jenny, two former circus elephants:** Jessica Sarter, "How an Isolated Elephant Found Love Again," *One Green Planet* "Animals and Nature," February 12, 2016, www.onegreenplanet.org, http://www.onegreenplanet.org/animalsandnature/how-an-isolated-elephant-found-love-again-shirleys-story/.

29 **Cows tend to form close relationships:** K. M. McLennan, "Social bonds in dairy cattle: The effect of dynamic group systems on welfare and productivity" (PhD diss., University of Northampton, 2013).

29 **A group of mature female humpback whales:** Matt Walker, "Humpback whales form friendships that last years," *BBC* "Earth News," June 7, 2010, www.bbc.co.uk, http://news.bbc.co.uk/earth/hi/earth_news/newsid_8722000/8722626.stm; Christian Ramp, Wilhelm Hagen, Per

Palsbøll, et al. "Age-related multi-year associations in female humpback whales (*Megaptera novaeangliae*)," *Behavorial Ecology and Sociobiology* 64, no. 10 (October 2010): 1563–76, doi:10.1007/s00265–010–0970–8.

29 **In 2006, two scientists:** Patrick R. Hof, and Estel Van Der Gucht, "Structure of the cerebral cortex of the humpback whale, *Megaptera novaeangliae* (Cetacea, Mysticeti, Balaenopteridae)," *The Anatomical Record* 290, no. 1 (January 23, 2007): 1–31, doi:10.1002/ar.20407.

30 **Rosing was in Churchill, Manitoba:** "Animals at Play: A Talk by Stuart Brown," *On Being* vimeo video, 2:20, produced by Trent Gilliss, based on photographs by Norbert Rosing, https://vimeo.com/282517.

31 **Another surprising story:** Mark H. Deakos, Brian K. Branstetter, Lori Mazzuca, et al., "Two unusual interactions between a bottlenose dolphin (*Tursiops truncatus*) and a humpback whale (*Megaptera novaeangliae*) in Hawaiian waters," *Aquatic Mammals* 36, no. 2 (June 2010): 121–28.

31 **It is most often in captivity:** Jennifer S. Holland, *Unlikely Friendships: 47 Remarkable Stories from the Animal Kingdom* (New York: Workman Publishing, 2011), 7–9, 51–53, 65–67, 77–80.

32 **Such a friendship began:** Paul Kiernan, "Beached in Brazil, a Young Penguin Finds His Human Soul Mate," *Wall Street Journal* online, October 22, 2015, www.wsj.com, https://www.wsj.com/articles/beached-in-brazil-a-young-penguin-finds-his-human-soul-mate-1445560828.

Chapter 5. For the Fun of It: Play and Imagination

35 **Once, when paraglider Tim Hall:** Bernd Heinrich, *Mind of the Raven: Investigations and Adventures with Wolf-birds* (New York: Harper Collins, 2006), 291.

35 **Another example of ravens using an object:** John M. Marzluff and Tony Angell, *Gifts of the Crow: How Perception, Emotion, and Thought Allow Smart Birds to Behave Like Humans* (New York: Free Press, 2012), 117.

35 **Ravens have yet another in-flight game:** Bernard Heinrich and Rachael Smolker, "Play in common ravens (*Corvus corax*)," in *Animal Play: Evolutionary, Comparative and Ecological Perspectives*, eds. Marc Bekoff and John A. Byers (Cambridge, England: Cambridge University Press, 1998), 27–44.

35 **Play-fighting female paper wasps:** Sarah Zylinski, "Fun and play in invertebrates," *Current Biology* 25, no. 1 (January 5, 2015): R10-R12, doi:10.1016/j.cub.2014.09.068.

36 **Captive fish riding bubbles:** Jonathan Balcombe, *What a Fish Knows: The Inner Lives of Our Underwater Cousins* (New York: Scientific American/Farrar, Straus and Giroux, 2016), 94–98.

36 **Alligators and crocodiles play, too:** Vladimir Dinets, "Play Behavior in Crocodilians," *Animal Behavior and Cognition* 2, no. 1 (2015): 49–55, doi: 10.12966/abc.02.04.2015.

37 **He believes animals have fun when they play:** Jonathan P. Balcombe, *Pleasurable Kingdom: Animals and the Nature of Feeling Good* (London: Macmillan, 2006), 69.

37 **When it snows, prairie dogs:** Sarah Fecht, "Why Do Animals Love To Play In The Snow?," *Popular Science* online, January 25, 2016, www.popsci.com, http://www.popsci.com/why-do-animals-love-play-snow.

37 **Japanese macaques:** Kaitlyn Foley-Czap, "Japanese Snow Macaque," New England Primate Conservancy, www.neprimateconservancy.org, from Foley-Czap's blog "Letters from the Field" series, January 2011, http://www.neprimateconservancy.org/japanese-snow-monkey.html.

37 **In Alaska and Canada:** Heinrich and Smolker, "Play in common ravens (*Corvus corax*)," 27–44.

38 **In 2012, one Russian crow:** "Russian roof-surfin' bird caught on tape," YouTube video, 1:23, posted by "Green Acre," February 23, 2016, https://www.youtube.com/watch?v=UlvLWtBdB78.

38 **A musher who saw buffalo playing on ice:** Gary, Paulsen, *Winterdance: The Fine Madness of Running the Iditarod* (New York: Harcourt Brace, 1994), 193–94.

39 **Another species that dolphins "play" with:** Ben Wolford, "Do Dolphins Get High? BBC Cameras Catch Dolphins Chewing On Pufferfish Toxins," *International Science Times*, December 30, 2013, www.isciencetimes.com, http://www.iscience times.com/articles/6595/20131230/dolphins-high-bbc-cameras-catch-chewing-pufferfish.htm.

39 **One day while hanging out on the beach:** Dana McGregor in discussion with the author, November 2016. See http://surfinggoats.com/.

40 **What is known as *funktionslust*:** Balcombe, *Pleasurable Kingdom*, 17–18, 69.

40 **Pretending that sticks and small logs:** Sonya M. Kahlenberg and Richard W. Wrangham, "Sex differences in chimpanzees' use of sticks as play objects resemble those of children," *Current Biology* 20, no. 24 (December 21, 2010), doi:10.1016/j.cub.2010.11.024, http://www.cell.com/current-biology/pdf/S0960-9822(10)01449-1.pdf.

41 **Viki, a chimpanzee raised:** M. L. A. Jensvold and R. S. Fouts, "Imaginary play in chimpanzees (*Pan troglodytes*)," *Human Evolution* 8, no. 3 (July 1993): 217–27, doi:10.1007/BF02436716.

41 **Other captive chimpanzees:** Heidi Lyn, Patricia Greenfield, and Sue Savage-Rumbaugh, "The development of representational play in chimpanzees and bonobos: Evolutionary implications, pretense, and the role of interspecies

communication," *Cognitive Development* 21, no. 3 (July–September 2006): 199–213.

Chapter 6. Random Acts of Kindness: Empathy and Altruism

44 **To test this widespread belief:** Deborah M. Custance and Jennifer Mayer, "Empathic-like responding by domestic dogs (*Canis familiaris*) to distress in humans: An exploratory study," *Animal Cognition* 15, no. 5 (September 2012): 851–59, doi:10.1007/s10071–012–0510–1.

45 **Scientist Isabella Merola recruited:** I. Merola, M. Lazzaroni, S. Marshall-Pescini, et al., "Social referencing and cat–human communication," *Animal Cognition* 18, no. 3 (May 2015) 639–48, doi:10.1007/s10071–014–0832–2.

45 **In another study, researchers:** Moriah Galvan and Jennifer Vonk, "Man's other best friend: Domestic cats (*F. silvestris catus*) and their discrimination of human emotion cues," *Animal Cognition* 19, no.1 (January 2016): 193–205, doi:10.1007/s10071–015–0927–4.

45 **To test whether the prolific rat:** Inbal Ben-Ami Bartal, Jean Decety, and Peggy Mason, "Empathy and Pro-Social Behavior in Rats," *Science* 334, no. 6061 (December 9, 2011): 1427–30, doi:10.1126/science.1210789.

46 **Rats would rather eat chocolate:** Nobuya Sato, Ling Tan, Kazushi Tate, et al., "Rats demonstrate helping behavior toward a soaked conspecific," *Animal Cognition* 18, no. 5 (September 2015): 1039–47, doi:10.1007/s10071–015–0872–2.

46 **Ravens might have empathetic friendships:** Orlaith N. Fraser and Thomas Bugnyar, "Do Ravens Show Consolation? Responses to Distressed Others," *PLoS One* 5, no. 5 (May 12, 2010): e10605, doi:10.1371/journal.pone.0010605.

47 **As for fish, studies have established:** Culum Brown," Fish intelligence, sentience and ethics," *Animal Cognition* 18, no. 1 (2015): 1–17, doi:10.1007/s10071–014–0761–0, http://animalstudiesrepository.org/cgi/viewcontent.cgi?article=1074&context=acwp_asie.

47 **Consider the crow and the cat:** Linda Cole, "The True Story of a Wild Crow That Saved a Kitten," Canidae Pet Food Company (Blog), April 26, 2102, www.canidae.com, https://www.canidae.com/blog/2012/04/true-story-of-wild-crow-that-saved/; "Crow and Kitten Are Friends," YouTube video, 7:29, PAX TV Network, posted by "Ozricus" on June 2, 2007, https://www.youtube.com/watch?v=1JiJzqXxgxo.

48 **Frans de Waal tells of Kuni:** Frans B. M. de Waal, *Our Inner Ape: A Leading Primatologist Explains Why We Are Who We Are* (New York: Riverhead Books, 2005), Kindle edition, 2.

49 **Marine ecologist Robert Pitman saw:** Robert L. Pitman, Volker B. Deecke, Christine M. Gabriele, et al., "Humpback whales interfering when mammal-eating killer whales attack other species: Mobbing behavior and interspecific altruism?," *Marine Mammal Science* 33 (July 20, 2016): 7–58, doi:10.1111/mms.12343.

49 **A final story of interspecies empathy:** "Chantek, the first orangutan person | Lyn Miles |TEDxUT-Chattanooga," YouTube video, 16:27, posted by TEDx, Nov 17, 2014, https://www.youtube.com/watch?v=q2pisrdO2TQ.

chapter 7. a sense of the sacred: death and spirituality

51 **The story of Storm:** Barbara J. King, *How Animals Grieve* (Chicago: University of Chicago Press, 2013), 32–33.

51 **A red fox bury her mate:** Marc Bekoff, *The Emotional Lives of Animals: A Leading Scientist Explores Animal Joy, Sorrow, and Empathy—and Why They Matter* (Novato, CA: New World Library, 2007), 63–64.

51 **One day he observed four magpies:** Bekoff, *The Emotional Lives of Animals*, 1–2.

52 **A similar story of a dead crow:** Jennifer Ackerman, *The Genius of Birds* (New York: Penguin Press, 2016), 134.

52 **Koko, the gorilla who knows sign language:** Francine Patterson and Wendy Gordon, "The Case for the Personhood of Gorillas," in *The Great Ape Project*, eds. Paola Cavalieri and Peter Singer (New York: St. Martin's Griffin, 1993), 58–77.

53 **A video posted online in 2016:** Matt Walker, "Chimps filmed grieving for dead friend," *BBC* "Earth" video, 5:48, May 18, 2016, www.bbc.com, http://www.bbc.com/earth/story/20160517-chimps-grieve-for-dead-friend.

53 **The story of two goats—Myrtle and Blondie:** King, *How Animals Grieve*, 36–38.

54 **Wild elephants often stand vigil:** King, *How Animals Grieve*, 52–63.

54 **She has often told the story of the:** Jane Goodall, "Epilogue: The Dance of Awe," in *A Communion of Subjects: Animals in Religion, Science, and Ethics*, eds. Paul Waldau and Kimberley C. Patton (New York: Columbia University Press, 2006): 651–56.

55 **Fire has elicited a similar response:** James B. Harrod, "The Case for Chimpanzee Religion," *Journal for the Study of Religion, Nature and Culture* 8, no. 1 (2014): 27–33, doi:10.1558/jsrnc.v8i1.8; Jill D. Pruetz and Thomas C. LaDuke, "Brief Communication: Reaction to Fire by Savanna Chimpanzees (*Pan troglodytes verus*) at Fongoli, Senegal: Conceptualization of 'Fire Behavior' and the Case for a Chimpanzee Model," *American*

Journal of Physical Anthropology 141, no. 4 (April 1, 2010): 646–50, doi:10.1002/ajpa.21245.

55 **Chimpanzees behaving strangely toward:** Hjalmar S. Kühl, Ammie K. Kalan, Mimi Arandjelovic, et al., "Chimpanzee accumulative stone throwing," *Scientific Reports* 6, no. 22219 (February 29, 2016), doi:10.1038/srep22219.

55 **Chimps aren't the only primates:** Barbara Smuts, "Encounters with animal minds," *Journal of Consciousness Studies* 8, no. 5–7 (2001): 300–01, http://www2.warwick.ac.uk/fac/cross_fac/iatl/activities/modules/ugmodules/humananimalstudies/lectures/32/smuts.pdf.

57 **These and other ritualized behaviors:** Donovan O. Schaefer, *Religious Affects: Animality, Evolution, and Power* (Durham, NC: Duke University Press, 2015), 1–35.

CHAPTER 8. TO KNOW THYSELF: AWARENESS AND IDENTITY

61 **Once, after unsuccessfully begging:** Francine Patterson and Wendy Gordon, "The Case for the Personhood of Gorillas," in *The Great Ape Project*, eds. Paola Cavalieri and Peter Singer (New York: St. Martin's Griffin, 1993), 65.

61 **When asked what she sees in the mirror:** Patterson and Gordon, "The Case for the Personhood of Gorillas," 76.

61 **The renowned bonobo, Kanzi:** Paul Raffael, "Speaking Bonobo," *Smithsonian* online, November 2006, www.smithsonian.com, http://www.smithsonianmag.com/science-nature/speaking bonobo-134931541/; Sue Savage-Rumbaugh and Roger Lewin, *Kanzi: The Ape at the Brink of the Human Mind* (New York: Wiley, 1994), Kindle edition, 266.

61 **Kanzi used a video camera:** Savage-Rumbaugh and Lewin, *Kanzi: The Ape at the Brink of the Human Mind*, 266.

62 **Another ape, the orangutan Chantek:** H. L. Miles, "Me Chantek: The development of self-awareness in a signing orangutan," in *Self-Awareness in Animals and Humans: Developmental Perspectives*, eds. Sue Taylor Parker, Robert W. Mitchell, and Maria L. Boccia (New York: Cambridge University Press, 1994), 254–69.

62 **He pointed to the moon and asked:** Chris Herzfeld, *Wattana: An Orangutan in Paris*, trans. Oliver Y. Martin and Robert D. Martin (Chicago: University of Chicago Press, 2016), 63–64.

62 **Described himself as an "orangutan person":** Susanne Antonetta, "Language Garden," *Orion*, www.orionmagazine.org, https://orionmagazine.org/article/language-garden/; H. Lyn White Miles, "Can Chantek Talk in Codes?" in *Anthropology: The Human Challenge*, 15th ed., eds. William A. Haviland, Harald E. L. Prins, Dana Walrath, et al. (Boston: Cengage Learning, 2014, 2017), 112–13.

62 **A self-awareness assessment tool:** Gordon G. Gallup, Jr., James R. Anderson, and Daniel J. Shillito, "The Mirror Test," in *The Cognitive Animal: Empirical and Theoretical Perspectives on Animal Cognition*, eds. Marc Bekoff, Colin Allen, and Gordon M. Burhgardt (Cambridge: Massachusetts Institute of Technology, 2002), 325–33.

62 **To date, eleven species:** Amanda Pachniewska, "List of Animals That Have Passed the Mirror Test," Animal Cognition, www.animalcognition.org, http://www.animalcognition.org/2015/04/15/list-of-animals-that-have-passed-the-mirror-test/; Csilla Ari and Dominic P. D'Agostino, "Contingency checking and self-directed behaviors in giant manta rays: Do elasmobranchs have self-awareness?," *Journal of Ethology* 34, no. 2 (May 2016): 167–174, doi:10.1007/s10164–016–0462-z; Koji Toda and Michael L. Platt, "Animal Cognition: Monkeys Pass the Mirror Test," *Current Biology* 25, no. 2 (January 19, 2015): R64-R66, doi:10.1016/j.cub.2014.12.005.

63 **One chimpanzee, Austin, who lived at the:** Savage-Rumbaugh and Lewin, *Kanzi: The Ape at the Brink of the Human Mind*, Kindle edition, 265–66.

63 **European magpies were the first:** Helmut Prior, Ariane Schwarz, and Onur Güntürkün, "Mirror-Induced Behavior in the Magpie (Pica pica): Evidence of Self-Recognition," *PLoS Biology* 6, no. 8 (August 19, 2008): e202, doi:10.1371/journal.pbio.0060202.

63 **The manta ray:** Ari and D'Agostino, "Contingency checking and self-directed behaviors in giant manta rays," 167–174.

64 **Even an insect:** Marie-Claire Cammaerts and Roger Cammaerts, "Are Ants (*Hymenoptera, Formicidae*) Capable of Self Recognition?" *Journal of Science* 5, no. 7 (2015): 521–32, http://www.journalofscience.net/File_Folder/521-532(jos).pdf.

64 **So far the only gorilla to pass:** Patterson and Gordon, "The Case for the Personhood of Gorillas," 58.

64 **For example, in Kenya:** Maggie Koerth-Baker, "Kids (and Animals) Who Fail Classic Mirror Tests May Still Have Sense of Self," *Scientific American*, November 29, 2010, https://www.scientificamerican.com/article/kids-and-animals-who-fail-classic-mirror/.

65 **In an experiment Bekoff nick-named:** Marc Bekoff, "Observations of Scent-marking and Discriminating Self from Others by a Domestic Dog (*Canis familiaris*): Tales of Displaced Yellow Snow," *Behavioural Processes* 55, no. 2 (September 2001): 75–79, doi:10.1016/S0376–6357(01)00142–5; Roberto Cazzolla Gatti, "Self-consciousness: Beyond the looking-glass and what dogs found there," *Ethology, Ecology & Evolution* 28, no. 2 (November 2015): 232–40, doi:10.1080/03949370.2015.1102777.

65 **Koko the gorilla has been known to tell a lie:** Patterson and Gordon, "The Case for the Personhood of Gorillas," 58.

66 **But if they notice that other birds:** Edward W. Legg and Nicola S. Clayton, "Eurasian jays (*Garrulus glandarius*) conceal caches from onlookers," *Animal Cognition* 17, no. 5 (September 2014): 1223–26, doi:10.1007/s10071–014–0743–2.

66 **When a bottlenose dolphin is just a few months:** Stephanie L. King and Vincent M. Janik, "Bottlenose dolphins can use learned vocal labels to address each other," *Proceedings of the National Academy of Sciences* 110, no. 32 (July 22, 2013): 13216–21, doi:10.1073/pnas.1304459110; Stephanie L. King, Laela S. Sayigh, Randall S. Wells, et al., "Vocal copying of individually distinctive signature whistles in bottlenose dolphins," *Proceedings of the Royal Society B* 280 (February 20, 2013): https://doi.org/10.1098/rspb.2013.0053.

68 **Each parrot appears to have:** Karl S. Berg, Soraya Delgado, Kathryn A. Cortopassi, et al., "Vertical transmission of learned signatures in a wild parrot, *Proceedings of the Royal Society B* 279 (February 7, 2011): 585–91, doi:10.1098/rspb.2011.0932.

69 **Take Puck, for example:** "Puck: The Talking Wonder," *Incredible Budgerigar* (Blog), February 2013, www.incrediblebudgerigar.blogspot.com, http://incrediblebudgerigar.blogspot.com/2013/02/puck-talking-wonder.html.

Chapter 9. If We Could Talk to the Animals: Language

71 **If prairie dogs are starting to sound:** Con Slobodchikoff, *Chasing Doctor Dolittle: Learning the Language of Animals* (New York: St. Martin's Press, 2012); Con N. Slobodchikoff (2002) "Cognition and communication in Gunnison's prairie dogs," in *The Cognitive Animal: Empirical and Theoretical Perspectives on Animal Cognition*, eds. Marc Bekoff, Colin Allen, and Gordon M. Burhgardt (Cambridge: Massachusetts Institute of Technology, 2002), 257–64; C. N. Slobodchikoff, J. Kiriazis, C. Fischer, et al., "Semantic information distinguishing individual predators in the alarm calls of Gunnison's prairie dogs," *Animal Behaviour* 42, no. 5 (November 1991): 712–19, doi:10.1016/S0003–3472(05)80117–4; Con N. Slobodchikoff, Andrea Paseka, Jennifer L. Verdolin, "Prairie dog alarm calls encode labels about predator colors," *Animal Cognition* 12, no. 3 (May 2009): 435–39, doi:10.1007/s10071–008–0203-y.

73 **Slobodchikoff has decoded more than one hundred:** Con N. Slobodchikoff, email message to author, April 20, 2015.

75 **A different hypothesis about why the owls:** Rebecca D. Bryan and Michael B. Wunder, "Western Burrowing Owls (*Athene cunicularia*

hypugaea) Eavesdrop on Alarm Calls of Black-Tailed Prairie Dogs (*Cynomys ludovicianus*)," *Ethology* 120 (2014): 180–88, doi:10.1111/eth.12194; Cat Ferguson, "Squatting Owls Eavesdrop on Prairie Dogs," *Inside Science*, "Creature," www.insidescience.org, January 16, 2014, https:/www.insidescience.org/news/squatting-owls-eavesdrop-prairie-dogs.

76 **Captive orcas living with:** Whitney B. Musser, Ann E. Bowles, Dawn M. Grebner, et al., "Differences in acoustic features of vocalizations produced by killer whales cross-socialized with bottlenose dolphins," *Journal of the Acoustical Society of America* 136, no. 4 (October 2014): 1990–2002, doi:10.1121/1.4893906.

76 **Meanwhile, halfway around the world:** Dorothee Kremers, Margarita Briseño Jaramillo, M. Böye, et al., "Do dolphins rehearse show-stimuli when at rest? Delayed matching of auditory memory," *Frontiers in Psychology* 2 (December 29, 2011): 386, doi.org/10.3389/fpsyg.2011.00386; Bill Blakemore, "Dolphins Reported Talking Whale in Their Sleep," *ABC News* (Blog), January 29, 2012, www.abcnews.go.com, http://abcnews.go.com/blogs/technology/2012/01/dolphins-reported-talking-whale-in-their-sleep/.

77 **Pilley wanted to know just how large**: John W. Pilley and Alliston K. Reid, "Border collie comprehends object names as verbal referents," *Behavioural Processes* 86, no. 2 (February 2011): 184–95, doi:10.1016/j.beproc.2010.11.007.

78 **Tried to teach a chimpanzee named Viki:** Erin Wayman, "Six Talking Apes," *Smithsonian* online, August 11, 2011, www.smithsonian.com, http://www.smithsonianmag.com/science-nature/six-talking-apes-48085302/.

78 **Washoe was the first chimpanzee:** Wayman, "Six Talking Apes."

79 **Perhaps the most famous "talking ape":** Francine Patterson and Wendy Gordon, "The Case for the Personhood of Gorillas," in *The Great Ape Project*, eds. Paola Cavalieri and Peter Singer (New York: St. Martin's Griffin, 1993), 76.

79 **After experiencing an earthquake:** Patterson and Gordon, "The Case for the Personhood of Gorillas," 76.

79 **Chantek learned 150 signs:** William A. Hillix and Duane M. Rumbaugh, *Animal Bodies, Human Minds: Ape, Dolphin, and Parrot Language Skills* (New York: Kluwer Academic/Plenum Publishers, 2004), 194–99; "The Ape Who Went to College," YouTube video, 43:50, posted by Patrick Johnson, November 8, 2016, "The Ape Who Went to College," PBS, "My Wild Affair," 53:41, July 23, 2014.

79 **When he wanted a confidential exchange:** Susanne Antonetta, "Language Garden," *Orion*, www.orionmagazine.org, https://orionmagazine.org/article/language-garden/.

79 **Words could have multiple meanings:** H. Lyn White Miles, "Language and the Orangutan: The Old 'Person' of the Forest," in *The Great Ape Project*, eds. Paola Cavalieri and Peter Singer (New York: St. Martin's Griffin, 1993): 42–57.

79 **A sign language project with a chimp named:** Cindy Rodríguez, "Hanging from the Language Tree," *Columbia* (Fall 2009), http://www.columbia.edu/cu/alumni/Maga zine/Fall2009/feature4.html.

80 **In the early 1980s psychologist Sue:** Sue Savage-Rumbaugh and Roger Lewin, *Kanzi: The Ape at the Brink of the Human Mind* (New York: Wiley, 1994), Kindle edition; William A. Hillix and Duane M. Rumbaugh, *Animal Bodies, Human Minds*, 173–80.

81 **The "fairy tale" about talking animals**: Jane C. Hu, "What Do Talking Apes Really Tell Us?: The strange, disturbing world of Koko the gorilla and Kanzi the bonobo," *Slate* "Science," August 20, 2014, www.slate.com, http://www.slate.com/articles/health_and_science/science/2014/08/koko_kanzi_and_ape_language_research_criticism_of_working_conditions_and.html.

82 **Chimpanzees have their own:** Catherine Hobaiter and Richard W. Bryne, "The Meanings of Chimpanzee Gestures," *Current Biology* 24, no. 14 (July 21, 2014): 1596-1600, doi: 10.1016/j.cub.2014.05.066.

83 **Here are a few of the terms invented:** Tom Leonard, "He can cook, play music, use a computer—and make sarcastic jokes chatting with his 3,000-word vocabulary: My lunch with the world's cleverest chimp (who Skyped me later for another chat)," *Daily Mail*, June 7, 2014, www.dailymail.com, http://www.dailymail.co.uk/news/article-2651004/He-cook-play-music-use-computer-make-sarcastic-jokes-chatting-3—000-word-vocabulary-My-lunch-worlds-cleverest-chimp.html#ixzz4Vlu3Jiwg; Patterson and Gordon, "The Case for the Personhood of Gorillas," 65; H. Lyn White Miles, "Can Chantek Talk in Codes?" in *Anthropology: The Human Challenge*, 15th ed., eds. William A. Haviland, Harald E. L. Prins, Dana Walrath, et al. (Boston: Cengage Learning, 2014, 2017): 112–13; B. T. Gardner, R. A. Gardner, and S. G. Nichols, "The Shapes and Uses of Signs in a Cross-Fostering Laboratory," in *Teaching Sign Language to Chimpanzees*, eds. R. Allen Gardner, Beatrix T. Gardner, and Thomas E. Van Cantfort (Albany: State University of New York Press, 1989), 81.

Chapter 10. Count Them In: Numerical Cognition

86 **To test whether the bears could "count":** Jennifer Vonk and Michael J. Beran, "Bears 'Count' Too: Quantity Estimation and Comparison in Black Bears (*Ursus Americanus*)," *Animal Behaviour* 84, no. 1 (July 2012): 231–38, doi:10.1016/j.anbehav.2012.05.001.

87 **Nieder trained crows to peck:** Helen M. Ditz and Andreas Nieder, "Neurons selective to the number of visual items in the corvid songbird endbrain," *Proceedings of the National Academy of Sciences*, 112, no. 25 (June 23, 2015): 7827–32, doi:10.1073/pnas.1504245112, http://www.pnas.org/content/112/25/7827.full.pdf; "Crows count on 'number neurons,'" *ScienceDaily* "Science News," June 8, 2015, www.sciencedaily.com, based on the research of Helen M. Ditz and Andreas Nieder, Universitaet Tubingen, www.sciencedaily.com/releases/2015/06/150608152002.htm.

87 **As for chicks, within a few days of hatching:** Rosa Rugani, Laura Fontanari, Eleonora Simoni, et al., "Arithmetic in newborn chicks," *Proceedings of the Royal Society B* 276, no. 1666 (July 7, 2009): 2451–60, doi:10.1098/rspb.2009.0044.

88 **He could count up to eight:** Irene M. Pepperberg, *Alex & Me: How a Scientist and a Parrot Discovered a Hidden World of Animal Intelligence—and Formed a Deep Bond in the Process* (New York: HarperCollins, 2008); Irene M. Pepperberg, "Further evidence for addition and numerical competence by a Grey parrot (*Psittacus erithacus*)," *Animal Cognition* 15, no. 4 (July 2012): 711, doi:10.1007/s10071-012-0470-5.

89 **Able to understand and vocalize the abstract:** Irene M. Pepperberg, email exchange with author, January 8, 2016.

90 **Even the humble honeybee:** Hans J. Gross, Mario Pahl, Aung Si, et al. "Number-Based Visual Generalisation in the Honeybee," PLoS One 4, no.1 (January 28, 2009): e4263, doi:10.1371/journal.pone.0004263.

91 **Having some kind of numeric cognition:** Nellie Kamkar, "From Primates to Ants: Reviewing the Evidence of Numerical Cognition in Non-Human Species," *Western Undergraduate Psychology Journal*, 3, no. 1 (2015), http://ir.lib.uwo.ca/cgi/viewcontent.cgi?article=1032&context=wupj.

Chapter 11. Technology across the Kingdom: Tool Use

93 **In 2006, researchers found evidence:** Julio Mercader, Huw Barton, Jason Gillespie, et al., "4,300-Year-old chimpanzee sites and the origins of percussive stone technology," *Proceedings of the National Academy of Sciences* 104, no. 9 (February 2007): 3043–48, doi:10.1073/pnas.0607909104.

94 **Chimpanzee tool use in 1960:** David Quammen, "Being Jane Goodall," *National Geographic* online, October 2010, http://ngm.nationalgeographic.com/2010/10/jane-goodall/quammen-text.

94 **Chimpanzees use an impressive number of:** Robert W. Shumaker, Kristina R. Walkup, and Benjamin B. Beck, *Animal Tool Behavior: The Use and Manufacture of Tools by Animals*, rev. ed. (Baltimore: Johns Hopkins University Press,

2011); William C. McGrew, "Chimpanzee Technology," *Science* 328, no. 5978 (April 30 2010): 579–80, doi:10.1126/science.1187921.

95 **The only other animals that make stabbing:** J. D. Pruetz, P. Bertolani, K. Boyer Ontl, et al., "New evidence on the tool-assisted hunting exhibited by chimpanzees (*Pan troglodytes verus*) in a savannah habitat at Fongoli, Sénégal. *Royal Society Open Science* 2. no. 4 (April 15, 2015), doi:10.1098/rsos.140507.

96 **Fascinating examples of stone tool use:** Itai Roffman, Sue Savage-Rumbaugh, Elizabeth Rubert-Pugh, et al., "Stone tool production and utilization by bonobo-chimpanzees (*Pan paniscus*)," *Proceedings of the National Academy of Sciences* 109, no. 36 (July 2012): 14500–503, doi:10.1073/pnas.1212855109.

97 **Chimps have been imbibing a lot more than:** Kimberley J. Hockings, Nicola Bryson-Morrison, Susana Carvalho, et al., "Tools to tipple: Ethanol ingestion by wild chimpanzees using leaf-sponges," *Royal Society Open Science* 2, no. 6 (June 9, 2015), doi:10.1098/rsos.150150.

97 **Discovered only two species that make hooked:** Gavin R. Hunt and Russell D. Gray, "The crafting of hook tools by wild New Caledonian crows," *Proceedings of the Royal Society B* 271, S3 (February 7, 2004), doi:10.1098/rsbl.2003.0085; PDF version posted by United States National Center for Biotechonology Information at https://www.ncbi.nlm.nih.gov/pmc/articles/PMC1809970/pdf/15101428.pdf.

97 **A second distinctive tool: a serrated probe:** Hunt and Gray, "The crafting of hook tools by wild New Caledonian crows."

98 **These crows get attached to their tools:** Barbara C. Klump, Jessica E. M. van der Wal, James J. H. St Clair, et al., "Context-dependent 'safe-keeping' of foraging tools in New Caledonian crows," *Proceedings of the Royal Society B* 282, no. 1808 (May 20, 2015), doi.org/10.1098/rspb.2015.0278.

98 **Inspired by one of Aesop's fables:** Christopher David Bird and Nathan John Emery, "Rooks Use Stones to Raise the Water Level to Reach a Floating Worm," *Current Biology* 19, no. 16 (August 25, 2009): 1410–14, doi:10.1016/j.cub.2009.07.033.

98 **They began farming about fifty million years:** Susan Lumpkin and Stephanie Hsia, "The First Farmers," *Zoogoer* 33, no. 4 (July/August 2004), Smithsonian National Zoological Park, PDF version posted by University of Texas at Austin, Mathematics Department, at https://www.ma.utexas.edu/users/davis/375/LECTURES/L22/FONZ.pdf.

99 **These little cowboy ants capture aphids:** "Herding Aphids: How 'Farmer' Ants Keep Control of Their Food," *ScienceDaily* "Science News," October 11, 2007, www.sciencedaily.com, based on the research of Tom Oliver et

al., Imperial College London, www.sciencedaily.com/releases/2007/10/071009212548.htm.

Chapter 12. Finding Their Way: Spatial Intelligence

101 **Mauck and his colleagues had a hunch that:** Nele Gläser, Björn Mauck, Farid I. Kandil, et al., "Harbor Seals (*Phoca vitulina*) Can Perceive Optic Flow under Water," *PLoS One* 9, no. 7 (July 24, 2014): e103555, doi:10.1371/journal.pone.0103555.

102 **Whales might use stars as reference points:** Travis W. Horton, Richard N. Holdaway, Alexandre N. Zerbini, et al., "Straight as an arrow: Humpback whales swim constant course tracks during long-distance migration," *Biology Letters* 7, no. 5 (April 20, 2011): 674–79, doi:10.1098/rsbl.2011.0279.

104 **The beetle's dance is for orientation:** Emily Baird, Marcus J. Byrne, Jochen Smolka, et al., "The Dung Beetle Dance: An Orientation Behaviour?" *PLoS One* 7, no. 1 (January 18, 2012): e30211, doi:10.1371/journal.pone.0030211.

104 **To find out whether the beetles:** Marie Dacke, Emily Baird, Marcus Byrne, et al., "Dung Beetles Use the Milky Way for Orientation," *Current Biology* 23, no. 4 (February 2013): 298–300, doi:10.1016/j.cub.2012.12.034.

105 **A mental "snapshot" of the sky while doing:** Basil el Jundi, James J. Foster, Lana Khaldy, et al., "A Snapshot-Based Mechanism for Celestial Orientation," *Current Biology* 26, no. 11 (June 2016): 1456–62, doi:10.1016/j.cub.2016.03.030; "When dung beetles dance, they photograph the firmament," *ScienceDaily* "Science News," May 12, 2016, www.sciencedaily.com, based on the research of Marie Dacke et al., Lund University, Sweden, www.sciencedaily.com/releases/2016/05/160512125422.htm.

107 **So, just how do they accomplish this feat:** "Researcher Uncovering Mysteries of Memory By Studying Clever Bird," *ScienceDaily* "Science News," October 12, 2006, www.sciencedaily.com, based on the research of Brett Gibson, University of New Hampshire, www.sciencedaily.com/releases/2006/10/061012094818.htm.

107 **They use signposts:** Pavel Stopka and David W. Macdonald, "Way-marking behaviour: An aid to spatial navigation in the wood mouse (*Apodemus sylvaticus*)," *BMC Ecology* 3, no. 3 (April 2, 2003), doi:10.1186/1472-6785-3-3.

Chapter 13. For Art's Sake: Creativity and Aesthetics

109 **This artist was a chimpanzee:** Mark Hay, "Meet the Chimp Artist Who Outsold Renoir and Warhol," *Modern Notion* "History," October 7, 2014, www.modernnotion.com, http://modern-

notion.com/meet-the-chimp-artist-who-out-sold-renoir-and-warhol/.

110 **Both Koko the gorilla and her deceased gorilla:** Susan Thurston, "Gorillas as Artists," *Artists Ezine* 1, no. 6 (Spring/Summer 2006), http://www.artistsezine.com/WhyGorilla.htm.

110 **Wattana, a female orangutan:** Chris Herzfeld, *Wattana: An Orangutan in Paris*, trans. Oliver Y. Martin and Robert D. Martin (Chicago: University of Chicago Press, 2016), 85–111.

110 **Chicky, a dolphin living in captivity:** Lizzie Edmonds, "It's 'Chicasso'! Professional trainer teaches Chicky the dolphin how to PAINT," *Daily Mail*, October 21, 2013, www.dailymail.com, http://www.dailymail.co.uk/news/article-2470262/Dolphin-taught-PAINT-trainer.html.

111 **One elephant, Paya:** David Mikkelson, "Elephant Painting," Snopes.com, September 29, 2014, www.snopes.com, http://www.snopes.com/photos/animals/elephantpainting.asp; Desmond Morris, "Can jumbo elephants really paint? Intrigued by stories, naturalist Desmond Morris set out to find the truth," *Daily Mail*, 21 February 2009, www.dailymail.com, http://www.dailymail.co.uk/sciencetech/article-1151283/Can-jumbo-elephants-really-paint—Intrigued-stories-naturalist-Desmond-Morris-set-truth.html; "Elephant 'self-portrait' on show," *BBC News* online, July 21, 2006, www.news.bbc.co.uk, http://news.bbc.co.uk/2/hi/uk_news/scotland/edinburgh_and_east/5203120.stm.

112 **For millions of years, male bowerbirds:** David Rothenberg, *Survival of the Beautiful: Art, Science, and Evolution* (New York: Bloomsbury Press, 2011), 2.

112 **These artistic bowers are built:** Rothenberg, *Survival of the Beautiful*, 1–16; John A. Endler, "Bowerbirds, art and aesthetics: Are bowerbirds artists and do they have an aesthetic sense?," *Communicative & Integrative Biology* 5, no. 3 (May 2012): 281–83, doi:10.4161/cib.19481.

113 **Jam with musician Peter Gabriel:** "Peter Gabriel," Art for Bonobo Hope, www.artforbonobohope.org, http://artforbonobohope.org/project/peter-gabriel/; "Interspecies Internet 1," YouTube video, 2:17, posted by Peter Gabriel on July 2, 2013, https://www.youtube.com/watch?v=MM9YTlF_ErY; "Interspecies Internet 2," YouTube video, 0:43, posted by Peter Gabriel on July 2, 2013, https://www.youtube.com/watch?v=5iX92QRybGQ.

113 **Koko the gorilla got to play bass:** Bakkila Blake, "'This is the Day That I Will Never Forget': Flea Has a Jam Session with Koko the Gorilla," *People* online, August 21, 2016, www.people.com, http://people.com/celebrity/red-hot-chili-peppers-guitarist-flea-plays-guitar-with-koko-the-gorilla/.

114 **David Rothenberg had a different hypothesis:** David Rothenberg, Tina C. Roeske, Henning U. Voss, et al., "Investigation of musicality in birdsong," *Hearing Research* 308 (February 2014): 71–83, doi:10.1016/j.heares.2013.08.016; *Why*

Birds Sing, video, produced by BBC, available online at http://topdocumentaryfilms.com/why-birds-sing/.

114 **Humpback whales—the great musicians of the:** Linda N. Guinee and Katharine B. Payne, "Rhyme-like Repetitions in Songs of Humpback Whales," *Ethology* 79, no. 4 (December 1988): 295–306, doi:10.1111/j.1439-0310.1988.tb00718.x.

115 **Patel designed an experiment:** Robert Krulwich, "The List Of Animals Who Can Truly, Really Dance Is Very Short. Who's On It?" *NPR 88.5 WFDD*, "Krulwich Wonders: Robert Krulwich on Science," www.npr.org, April 1, 2014; "Snowball™—Our Dancing Cockatoo," YouTube video, 4:28, posted by BirdLoversOnly on October 15, 2007, https://www.youtube.com/watch?v=N7IZmRnAo6s.

115 **A sea lion named Ronan:** Peter Cook, Andrew Rouse, Margaret Wilson, et al., "A California sea lion (*Zalophus californianus*) can keep the beat: Motor entrainment to rhythmic auditory stimuli in a non vocal mimic, *Journal of Comparative Psychology* 127, no. 4 (November 2013): 412–27, doi:10.1037/a0032345.

115 **A video from the Pairi Daiza zoo:** "Are these elephants really dancing to classical music?" *The Guardian* (US Edition) video, 1:59, June 16, 2015, www.theguardian.com, sourced from International Television Network, https://www.theguardian.com/world/video/2015/jun/16/elephants-dancing-classical-music-video.

116 **This Celebes crested macaque:** David McAfee, "Copyright Office Says It Will Not Register 'Monkey Selfie'" August 22, 2014, Law 360, www.law360.com, https://www.law360.com/articles/570432/copyright-office-says-it-will-not-register-monkey-selfie.

117 **Capable of theatrical creativity:** Carl Safina, *Beyond Words: What Animals Think and Feel* (New York: Henry Holt and Company, 2015), 350.

Chapter 14. Re-imagining IQ: Thinking Outside the Human Brain

120 **He has learned to make the most of his many:** Jennifer Mather, Roland C. Anderson, and James B. Wood, *Octopus: The Ocean's Intelligent Invertebrate* (Portland, OR: Timber Press: 2010).

120 **This was what happened in 2016:** Karin Brulliard, "Octopus slips out of aquarium tank, crawls across floor, escapes down pipe to ocean," *Washington Post* online, April 13, 2016, https://www.washingtonpost.com/news/animalia/wp/2016/04/13/octopus-slips-out-of-aquarium-tank-crawls-across-floor-escapes-down-pipe-to-ocean/?utm_term=.30616242df77.

120 **Lucky Sucker, a two-spot Octopus:** Wendy Williams, *Kraken: The Curious, Exciting, and*

Slightly Disturbing Science of Squid (New York: Abrams Image, 2011), Kindle edition, 170.

121 **His friend found the octopus in his library:** Jacques Cousteau and Philippe Diolé, *Octopus and Squid: The Soft Intelligence* (Garden City, NY: Doubleday, 1973), 29.

121 **The octopus is also a tool user:** Robert W. Shumaker, Kristina R. Walkup, and Benjamin B. Beck, *Animal Tool Behavior: The Use and Manufacture of Tools by Animals*, rev. ed. (Baltimore: Johns Hopkins University Press, 2011), 30–31.

121 **The veined octopus collects:** Julian K. Finn, Tom Tregenza, and Mark D. Norman, "Defensive tool use in a coconut-carrying octopus," *Current Biology* 19, no. 23 (December 15, 2009): R1069–R1070, http://www.cell.com/current-biology/fulltext/S0960-9822(09)01914-9, doi:10.1016/j.cub.2009.10.052.

121 **Most interesting example of octopus tool use:** Matt Simon, "Absurd Creature of the Week: The Octopus That's Pretty Much Just a Swimming Blanket," *Wired* "Science," March, 20, 2015, www.wired.com, http://www.wired.com/2015/03/absurd-creature-week-blanket-octopus/.

121 **Personality test for octopuses:** Jennifer Mather, "Octopuses are Smart Suckers!?" *The Cephalopod Page*, www.the cephalopodpage.org, 2000, originally posted on www.manandmollusc.net, http://www.thecephalopodpage.org/smarts.php; Jennifer A. Mather and R.C. Anderson, "Personalities of octopuses (*Octopus rubescens*)," *Journal of Comparative Psychology* 107, no. 3 (January 1993): 336–40.

122 **The skin of the California two-spot octopus:** Julie Cohen, "Seeing Without Eyes," The *UC Santa Barbara Current*, "Science and Technology News" online, May 20, 2015, www.news.ucsb.edu, http://www.news.ucsb.edu/2015/015408/seeing-without-eyes.

122 **How he's going to camouflage:** James Wood and Kelsie Jackson, "How Cephalopods Change Color," *The Cephalopod Page*, www.the cephalopodpage.org, September, 16, 2004, PDF for the online Cephalopod School "Cephalopod Lessons Plans," http://www.thecephalopodpage.org/cephschool/HowCephalopodsChangeColor.pdf.

122 **And then there's texture:** Katherine Harmon Courage, "How the Octopus Creates Instant 3-D Camouflage On Its Skin," *Scientific American* (Blog), February 10, 2014, www.scientificamerican.com, https://blogs.scientificamerican.com/octopus-chronicles/how-the-octopus-creates-instant-3-d-camouflage-on-its-skin/.

123 **Scientists have observed the mimic octopus:** Sarah Jane Alger, "The Mimic Octopus: Master of Disguise," *Nature* "Scitable" (Blog), October 28, 2013, www.nature.com, http://www.nature.com/scitable/blog/accumulating-glitches/the_mimic_octopus_master_of

123 **Each of the two competitors watched a:** Sana Inoue and Tetsuro Matsuzawa, "Working memory of numerals in chimpanzees," *Current Biology* 17, no. 23 (December 4, 2007): R1004–R1005, doi:10.1016/j.cub.2007.10.027, http://www.cell.com/current-biology/fulltext/S0960-9822(07)02088-X "Chimp vs human!—Working Memory test," YouTube video, 3:42, *BBC Earth* "Extraordinary Animals—Earth," December 25, 2013, www.bbc.com/earth/world, https://www.youtube.com/watch?v=zsXP8qeFF6A; "Are you smarter than a chimp?," YouTube video, 1:43, posted by "Chimponaughty" on August 12, 2012, https://www.youtube.com/watch?v=qyJomdyjyvM.

124 **Biologist Colleen Reichmuth taught Rio:** C. R. Kastak, R. J. Schusterman, and D. Kastak, "Equivalence classification by California sea lions using class-specific reinforcers," *Journal of the Experimental Analysis of Behavior* 76, no. 2 (September 2001): 131–58, doi:10.1901/jeab.2001.76–131, https://www.ncbi.nlm.nih.gov/pmc/articles/PMC1284831/PMC1284831/; "Sea Lion Plays Snap!—Part 1," YouTube video, 2:21, *BBC Earth* "Extraordinary Animals—Earth," October 18, 2013, www.bbc.com/earth/world, https://www.youtube.com/watch?v=Ex1nEuzAYxo; "Clever Sea Lion Identifies Numbers and Letters!—Part 2," YouTube video, 3:16, *BBC Earth* "Extraordinary Animals—Earth," October 23, 2013, http://www.bbc.com/earth/world, https://www.youtube.com/watch?v=8yN_BBIE9fA.

125 **Then one of the dolphins, named Kelly:** Carl Safina, *Beyond Words: What Animals Think and Feel* (New York: Henry Holt and Company, 2015), 338.

126 **If an octopus made an intelligence test for:** James Wood, email message to author, May 24, 2016.

127 **So when octopus researchers put a crayfish:** F.B.M. de Waal, *Are We Smart Enough To Know How Smart Animals Are?* (New York: W.W. Norton & Company, 2016), Kindle edition, location 3694; R. C. Anderson and J. A. Mather, "It's All in the Cues: Octopuses (*Enteroctopus dofleini*) Learn to Open Jars," *Ferrantia: Proceedings of the 3rd international symposium: Coleoid Cephalopods Though Time* 59 (2010): 8–13, https://ps.mnhn.lu/ferrantia/publications/Ferrantia59.pdf.

127 **Werner Heisenberg once said:** Werner Heisenberg, *Physics and Philosophy: The Revolution in Modern Science* (New York: Harper, 1958): 78.

SUGGESTED READING

Ackerman, Jennifer. *The Genius of Birds*. New York: Penguin Press, 2016.

Balcombe, Jonathan P. *Pleasurable Kingdom: Animals and the Nature of Feeling Good*. London: Macmillan, 2006.

Balcombe, Jonathan P. *Second Nature: The Inner Lives of Animals*. Basingstoke: Palgrave Macmillan, 2011.

Balcombe, Jonathan P. *The Exultant Ark: A Pictorial Tour of Animal Pleasure*. Berkeley, CA: University of California Press, 2011.

Balcombe, Jonathan P. *What a Fish Knows: The Inner Lives of Our Underwater Cousins*. New York: Scientific American/Farrar, Straus and Giroux, 2016.

Bekoff, Marc, Colin Allen, and Gordon M. Burghardt, eds. *The Cognitive Animal: Empirical and Theoretical Perspectives on Animal Cognition*. Cambridge, MA: MIT Press, 2002.

Bekoff, Marc. *Minding Animals: Awareness, Emotions, and Heart*. New York: Oxford University Press, 2002.

Bekoff, Marc. *Animal Passions and Beastly Virtues: Reflections on Redecorating Nature*. Philadelphia: Temple University Press, 2006.

Bekoff, Marc. *The Emotional Lives of Animals: A Leading Scientist Explores Animal Joy, Sorrow, and Empathy—and Why They Matter*. Novato, CA: New World Library, 2007.

Bekoff, Marc, and Jessica Pierce. *Wild Justice: The Moral Lives of Animals*. Chicago: University of Chicago Press, 2009.

Boysen, Sarah T., and E. John. Capaldi, eds. *The Development of Numerical Competence: Animal and Human Models*. Hillsdale, NJ: L. Erlbaum Associates, 1993.

Dagg, Anne Innis. *Animal Friendships*. New York: Cambridge University Press, 2011.

de Waal, F.B.M. *The Age of Empathy: Nature's Lessons for a Kinder Society*. New York: Harmony Books, 2009.

de Waal, F.B.M. *The Bonobo and the Atheist: In Search of Humanism among the Primates*. New York: W. W. Norton & Company, 2013.

de Waal, F.B.M. *Are We Smart Enough to Know How Smart Animals Are?* New York: W. W. Norton & Company, 2016.

Emery, Nathan, and Frans de Waal. *Bird Brain: An Exploration of Avian Intelligence*. Princeton, NJ: Princeton University Press, 2016.

Heinrich, Bernd. *The Geese of Beaver Bog*. New York: ECCO, 2004.

Heinrich, Bernd. *Mind of the Raven: Investigations and Adventures with Wolf-birds*. New York: Harper Collins, 2006.

Heinrich, Bernd. *One Wild Bird at a Time: Portraits of Individual Lives*. Boston: Houghton Mifflin Harcourt, 2016.

Herzfeld, Chris. *Wattana: An Orangutan in Paris.* Translated by Oliver Y. Martin and Robert D. Martin. Chicago: University of Chicago Press, 2016.

Hillix, William A., and Duane M. Rumbaugh. *Animal Bodies, Human Minds: Ape, Dolphin, and Parrot Language Skills*. New York: Kluwer Academic/Plenum Publishers, 2004.

Holland, Jennifer S. *Unlikely Friendships: 47 Remarkable Stories from the Animal Kingdom*. New York: Workman Publishing, 2011.

King, Barbara J. *How Animals Grieve*. Chicago: University of Chicago Press, 2013.

Marzluff, John, and Tony Angell. *In the Company of Crows and Ravens*. New Haven: Yale University Press, 2005.

Marzluff, John, and Tony Angell. *Gifts of the Crow: How Perception, Emotion, and Thought Allow Smart Birds to Behave like Humans*. New York: Free Press, 2012.

Masson, Jefrey Moussaieff, and Susan McCarthy. *When Elephants Weep: The Emotional Lives of Animals.* New York: Dell Publishing, 1995.

Masson, Jeffrey Moussaieff. *The Pig Who Sang to the Moon: The Emotional World of Farm Animals.* New York: Ballantine Books, 2003.

Masson, Jeffrey Moussaieff. *Beasts: What Animals Can Teach Us about the Origins of Good and Evil.* New York: Bloomsbury, 2014.

Morell, Virginia. *Animal Wise: The Thoughts and Emotions of Our Fellow Creatures*. New York: Crown Publishers, 2013.

Parker, Sue Taylor, Robert W. Mitchell, and Maria L. Boccia, eds. *Self-awareness in Animals and Humans: Developmental Perspectives*. Cambridge: Cambridge University Press, 1994.

Pepperberg, Irene M. *Alex & Me: How a Scientist and a Parrot Discovered a Hidden World of Animal Intelligence—and Formed a Deep Bond in the Process*. New York: HarperCollins, 2008.

Peterson, Dale. *The Moral Lives of Animals*. New York: Bloomsbury Press, 2011.

Rothenberg, David. *Why Birds Sing: A Journey through the Mystery of Bird Song*. New York: Basic Books, 2005.

Rothenberg, David. *Thousand Mile Song: Whale Music in a Sea of Sound*. New York: Basic Books, 2008.

Rothenberg, David. *Survival of the Beautiful: Art, Science, and Evolution*. New York: Bloomsbury Press, 2011.

Safina, Carl. *Beyond Words: What Animals Think and Feel*. New York: Henry Holt and Company, 2015.

Schaefer, Donovan O. *Religious Affects: Animality, Evolution, and Power*. Durham: Duke University Press, 2015.

Shumaker, Robert W., Kristina R. Walkup, and Benjamin B. Beck. *Animal Tool Behavior: The Use and Manufacture of Tools by Animals*. rev. ed. Baltimore: Johns Hopkins University Press, 2011.

Waldau, Paul, and Kimberley Patton, eds. *A Communion of Subjects: Animals in Religion, Science, and Ethics*. New York: Columbia University Press, 2006.

INDEX